PREFACE

The International Energy Agency and its 21 Members pursue energy objectives through co-ordination and co-operation. The main goals are transparency of energy markets, energy efficiency, diversification of energy sources and the development of new technologies. The Agency's work assigns high priority to examining energy developments in other countries and assessing evolving global energy patterns. In order to have more flexible and less constrained energy economies, it is important that energy sectors and policies in the world at large are taken in an overall perspective. Appreciation of the oft-cited "interdependence" is thus mandatory for energy planners everywhere.

In this context, we have been developing energy balances for a number of non-OECD countries, and we have consulted with some of these countries as we and they work to upgrade our energy statistical systems. This work will permit us in the near future to publish approximately 70 energy balances for non-member countries, as well as a handbook which we hope will be of general use in aiding energy statisticians everywhere in collecting, collating, and presenting data in a meaningful, user-friendly manner.

We have also prepared, and present herewith, a series of reports treating a range of global energy topics. The issues presented in these papers have all been the subject of recent examination at the IEA. We hope their publication will provide others with some useful insights into a number of key aspects of the world energy situation.

The Secretariat was assisted in the preparation of this document by officials of Member governments and other experts. I am most grateful to all of them for their help without which the work could not have been completed. The work is, however, published on my responsibility as Executive Director of the IEA and does not necessarily reflect the views or positions of the IEA or its Member countries.

Helga Steeg
Executive Director

INTERNATIONAL ENERGY AGENCY

Energy
in non-OECD
Countries

s 1988

TION AND DEVELOPMENT

INTERNATIONAL ENERGY AGENCY

2, RUE ANDRÉ-PASCAL 75775 PARIS CEDEX 16, FRANCE

The International Energy Agency (IEA) is an autonomous body which was established in November 1974 within the framework of the Organisation for Economic Co-operation and Development (OECD) to implement an international energy programme.

It carries out a comprehensive programme of energy co-operation among twenty-one* of the OECD's twenty-four Member countries. The basic aims of IEA are:

i) co-operation among IEA Participating Countries to reduce excessive dependence on oil through energy conservation, development of alternative energy sources and energy research and development;

ii) an information system on the international oil market as well as consultation with oil companies;

iii) co-operation with oil producing and other oil consuming countries with a view to developing a stable international energy trade as well as the rational management and use of world energy resources in the interest of all countries;

iv) a plan to prepare Participating Countries against the risk of a major disruption of oil supplies and to share available oil in the event of an emergency.

*IEA Member countries: Australia, Austria, Belgium, Canada, Denmark, Germany, Greece, Ireland, Italy, Japan, Luxembourg, the Netherlands, New Zealand, Norway, Portugal, Spain, Sweden, Switzerland, Turkey, United Kingdom, United States.

Pursuant to article 1 of the Convention signed in Paris on 14th December, 1960, and which came into force on 30th September, 1961, the Organisation for Economic Co-operation and Development (OECD) shall promote policies designed:

– to achieve the highest sustainable economic growth and employment and a rising standard of living in Member countries, while maintaining financial stability, and thus to contribute to the development of the world economy;

– to contribute to sound economic expansion in Member as well as non-member countries in the process of economic development; and

– to contribute to the expansion of world trade on a multilateral, non-discriminatory basis in accordance with international obligations.

The original Member countries of the OECD are Austria, Belgium, Canada, Denmark, France, the Federal Republic of Germany, Greece, Iceland, Ireland, Italy, Luxembourg, the Netherlands, Norway, Portugal, Spain, Sweden, Switzerland, Turkey, the United Kingdom and the United States. The following countries became Members subsequently through accession at the dates indicated hereafter: Japan (28th April, 1964), Finland (28th January, 1969), Australia (7th June, 1971) and New Zealand (29th May, 1973).

The Socialist Federal Republic of Yugoslavia takes part in some of the work of the OECD (agreement of 28th October, 1961).

Cover photograph : IMAGE BANK.

TABLE OF CONTENTS

PART I

PART II

PART I

OVERVIEW

I. INTRODUCTION

The availability and use of energy resources has always been a concern since all forms of human activity — from cooking to transport to industry — require some form of energy. Modern economies have become very dependent on reliable and diverse supplies of energy to satisfy different and changing needs. While some countries are endowed with a relatively abundant resource base, others are heavily dependent on imports to meet their needs. Regardless, energy does not just appear: it has to be provided and it has a cost — often a high cost. This is true for imported energy; but it is also true for the domestic producer who must lay out large sums of capital for building hydro-electric dams, exploiting coal or oil reserves, gathering wood or developing transmission/transportation systems. It is also true for the consumer who must develop infrastructure to use primary energy directly or transform it to more usable forms. All of these actions require skills and specialized knowledge. For many countries, these costs and the access to these skills and technologies make the development of an efficient and economic energy system very difficult.

Since the first major oil price increase in 1973, governments have become much more aware of how crucial energy is to our economies and how important it is to have well-expressed and defined energy policies.

Primarily because of the two oil price increases in the 1970s but also because of the large costs in providing energy services, attitudes have changed. The main lesson learned since 1973 is that energy cannot be take for granted — economies, whether large or small, cannot afford to be without adequate supplies of energy. This has meant that governments, as well as energy industries, have had to become more active and determined in pursuing energy objectives to reduce vulnerability and maintain flexible, responsive and reliable energy systems. This is equally true for industrialized, centrally planned and developing economies.

The member countries of the International Energy Agency (IEA) have been pursuing common energy objectives since its formation in 1974 to reduce dependence on vulnerable energy sources. These have been in the form of improving energy security through an emergency oil sharing system, adequate oil stocks, diversification through increased indigenous production and fuel switching, conservation and R&D. Because of the importance of energy trade and growing global interdependence, the IEA also has a direct interest in the energy systems, the production and consumption prospects, and the major issues affecting energy developments in non-OECD countries. The major oil price increases in the 1970s showed the need for continuing international co-operation with respect to energy issues and monitoring of global energy flows.

Since the fall in the price of oil in early 1986, there has been a renewed need to assess the the adequacy of current energy policies and the trends in production and consumption. While the downturn in energy prices has been beneficial to consumers, it has caused severe economic problems for countries or regions dependent on oil production. For consuming countries the lower oil prices have affected trends towards further fuel diversification and energy conservation. There has been added concern about the viability of renewable sources of energy.

The IEA, itself, has been analysing these issues for its Member countries. In May 1987, IEA Ministers concluded that although policies since 1974 have been successful, there is a need for energy policies for the 1990s which will:

— maintain energy security through continued development of indigenous energy resources and technologies and improvements in the efficiency of energy use;

— secure the benefits for IEA countries as a whole, of lower energy and oil prices;

— promote free and open trade in energy; and

— further improve preparedness to deal with a disruption in energy supplies.

This report is an attempt to highlight some of the current energy issues and present some non-OECD energy data and production and consumption indicators that shed light on the current situation and may give an insight into the future. The report does not endeavour to be comprehensive nor prescriptive. Because it is not instructive to group all non-OECD countries together, they have been grouped in two ways: either regionally or by type of economy. The regions used are Latin America, Africa, Middle East, Asia-Pacific, and Centrally Planned Economies (including China). Otherwise they have been grouped as Oil Exporting Developing Countries (OXDC), Oil Importing Developing Countries — substantial oil producers (OIDC - substantial producers), other Oil Importing Developing Countries (OIDC - other) and Centrally Planned Economies (CPEs).

II. ENERGY AND THE ECONOMY

Energy production, consumption and trade have a very significant impact on many aspects of national economies — such as national income, balance of payments, employment, capital investment, external debt.

All economies produce and distribute some forms of energy (particularly electricity), some using indigenous energy sources and some importing them. Most countries rely on a mixture of the two. For some economies, energy is almost solely a means to provide a service to other parts of the society (heating homes, fueling factories, etc.). Nevertheless, they still need an infrastructure to produce and distribute electricity; refine and/or import and distribute petroleum products; obtain and distribute others forms of energy that are used nationally. For other countries the energy sector is in itself a means to expand the economy (oil, gas and coal exports, for example). Even those countries, however, have to provide useful energy for domestic needs.

The major oil price increases in the 1970s had opposite effects on importing and exporting countries. For example, for Brazil which has been a major oil importer, gross oil imports absorbed 12% of export revenues in 1973 92t this climbed to 50% in 1982. It later declined to 22% in 1984, largely due to a decrease in import requirements. Brazil's oil imports represented about 50% of import spending. This compares with interest on foreign debt taking 25%. As a reaction to high oil oil prices, Brazil decided to introduce a large alcohol fuels programme in an attempt to reduce oil imports and the strain on the balance of payments. In general, Table 1 shows the net oil export costs for developing countries and how it has gradually improved since 1981. This has been particularly true for the poorest countries.

For many oil exporting countries, oil revenue represents a very large share of national income. For example, petroleum represented 93% of merchandise exports in Saudi Arabia in 1983. Understandably, changes in energy exports and/or prices can have a serious effect on the economy, including balance of payments, debt charges, employment and national income.

Table 1 shows the net oil export earnings for the principal oil exporting countries since 1983. Oil revenues fell by 46% in 1986 compared to 1985.

III. TRENDS IN NON-OECD ENERGY PRODUCTION AND CONSUMPTION

Since the early 1970s, when oil and other energy prices started a steady rise, all regions have changed their pattern of energy use. This can be seen by Figure 1 which shows the shares of commercially traded primary energy. Biomass, which is the dominant fuel in many countries, has not been included in these calculations, largely because of the difficulty in developing the statistics. This is not to underplay the importance of biomass - which is generally in the form of fuelwood. It has been estimated that about 2 Mmt of fuelwood and charcoal are consumed daily in developing countries. Non-commercial fuels represent about 28% of TPER in India, all of this used in the residential sector. In 60-70 developing countries, it has been estimated that biomass represents over 50% of TPER.

Table 1 **Net Oil Export Earnings of Major Oil Exporting**
Developing Countries 1983 - 1986
($ billion)

	1983	1984	1985	1986
Algeria*	9.00	8.19	7.81	3.55
Angola	1.72	1.10	2.25	1.30
Bahrain	0.22	0.18	0.17	0.08
Brunei	1.79	1.60	1.52	0.84
Cameroon	1.15	1.25	1.70	0.86
Colombia	n.a.	n.a.	0.03	0.58
Congo	1.04	1.16	1.26	0.65
Ecuador*	1.55	1.71	1.94	0.89
Egypt	3.66	4.56	4.65	2.00
Gabon*	1.63	1.61	1.51	0.71
Indonesia*	10.28	11.24	9.04	4.64
Iran*	18.37	14.70	15.07	5.69
Iraq*	8.43	9.73	11.26	6.86
Kuwait*	7.91	8.50	7.86	5.28
Libya*	11.10	10.30	9.39	4.25
Malaysia	1.95	2.53	2.13	1.38
Mexico	15.51	15.89	14.06	5.70
Nigeria*	11.46	12.75	13.22	6.42
Oman	4.04	4.09	4.76	2.49
Peru	0.63	0.61	0.80	0.32
Qatar*	3.29	4.29	3.07	1.74
Saudi Arabia*	47.22	40.41	27.55	19.55
Syria	0.21	0.22	0.13	0.10
Trinidad	1.44	1.42	1.49	0.64
Tunisia	0.62	0.61	0.64	0.24
UAE*	12.02	13.72	12.27	7.27
Venezuela*	15.28	15.63	13.59	6.67
Total	191.52	188.00	169.17	90.74
% OPEC	82	81	79	81

*   OPEC Members.

Source: Secretariat estimates.

While oil is still the dominant fuel, in most areas other fuels have made
large inroads as substitutes for oil. Most noticeably, use of nuclear power
has emerged in the CPEs and the Asia-Pacific (primarily), and natural
gas and coal have made large gains in specific regions where economic
deposits of these fuels lie. All regions show quite different patterns of
change, depending on locally-available energy sources and the efforts
made to reduce dependence on oil.

Between 1975 and 1986, primary energy requirements increased 2.2%
per year in the CPEs, 4% per year for Latin America, 5% per year for

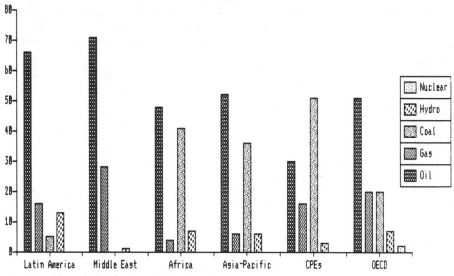

Figure 1a **Share of Primary Energy — 1975**
(%)

Source: BP Statistical Review of World Energy.

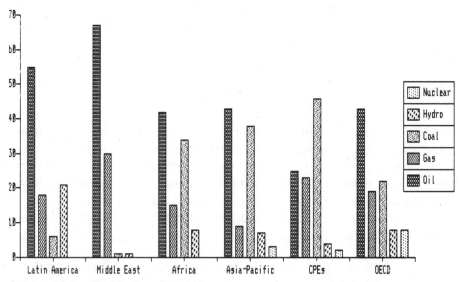

Figure 1b **Share of Primary Energy — 1986**
(%)

Source: BP Statistical Review of World Energy.

the Middle East, 5.9% for the Asia-Pacific region and 6.3% for Africa. In comparison, primary energy requirements grew at a much slower rate of 1.0% per year for OECD countries. Between 1985 and 1986, primary energy requirements grew for developing countries 2.9% and 3.3% for CPE's; for the OECD, TPER grew by 0.5% for that same year. It should be noted, however, that these percentage changes represent growth from very different size bases in terms of absolute volume of consumption. Africa, for example, has a much smaller base than Latin America and all developing countries have a smaller base than does the OECD.

Oil

Production

Between 1979 and 1987, non-OECD oil production has decreased an average of 1.9% per year from 51.4 mbd in 1979 to 44.2 mbd in 1986, even though CPEs and non-OPEC developing countries experience growth in production during that period. Between 1985 and 1986 total oil production rebounded, growing 6.4%, primarily as a result of a strong increase in production in non-OPEC developing countries and the growth continued between 1986 and 1987, although at a slower rate of 1.8%. Between 1986 and 1987 all regions except the OECD experienced growth in oil production.

Table 2 **Oil Production* 1979-87**
(mbd)

	1979	1986	1987	%Change 1987/86	%Change p.a. 1987/79
OPEC Countries	31.6	19.3	19.4	+0.5	−5.9
Non-OPEC Countries	5.3	8.2	8.9	+8.5	+6.7
Total Developing Countries	36.9	27.5	28.3	+2.9	−3.3
CPEs**	14.5	15.5	15.9	+1.9	+1.2
Total non-OECD	51.4	43.1	44.2	+2.6	−1.9
OECD	14.9	16.9	16.8	−0.6	+1.5
Total	66.3	59.9	61.0	+1.8	−1.0

* Includes NGLs, excludes non-conventional supplies.
** These data represent total CPE production.

Sources: IEA/OECD Energy Statistics; Secretariat Estimates.

Table 3 Oil Consumption 1979-87
(mbd)

	1979	1986	1987	%Change 1987/86	%Change p.a. 1987/79
Africa	1.38	1.83	1.89	3.3	4.0
Asia-Pacific	3.01	3.71	3.89	4.9	3.3
Middle East	1.50	2.52	2.55	1.2	3.8
Latin America	4.25	4.60	4.68	1.7	1.3
CPEs	13.20	13.40	13.50	0.8	0.3
Total non-OECD	23.70	26.06	26.51	1.7	1.4
OECD	41.60	35.20	35.70	1.4	−1.9
TOTAL	65.30	61.26	62.21	1.6	−0.6

Sources: IEA/OECD Energy Statistics; Secretariat Estimates.

Consumption

Between 1986 and 1987 oil consumption grew by about 1.7% for non-OECD countries (see Table 3). Growth was strongest in the Asia-Pacific and Africa while consumption grew only marginally in Africa. Factors such as ongoing industrialization and urbanisation, along with a growing need for transportation fuels, have been leading reasons for the increased consumption. Demographic growth — changes in population size, mix and economic status — have also affected oil consumption.

As shown in Table 4, there is a much higher use of heavy fuel oil in both developing countries and the Soviet Union than in the OECD. While heavy fuel oil use has fallen by 54% in OECD countries since 1978, it has been almost unchanged in both the Soviet Union and in developing

Table 4 **Product Share in Total Consumption - 1985**
(%)

	Developing Countries	USSR*	OECD
Motor Gasoline/Naphtha	21	17	36
Middle Distillates	41	30	35
Heavy Fuel Oil	28	42	14
Other Products	10	11	15
	100	100	100

* Data are not available for all CPEs.

Sources: IEA/OECD Energy Balances, Secretariat Estimates.

countries. Both the Soviet Union and the developing countries have relatively low consumption of motor gasoline, through less motorisation. On the other hand, middle distillate use is proportionately greater in developing countries, notwithstanding the lesser requirement for heating oil than in the Soviet Union and in many OECD countries.

Oil demand in OXDCs has grown much more rapidly than that of OIDCs:

Table 5 **Developing Country Oil Demand**
(mbd)

	1970	1973	1974	1982	1986	1987	% Change Per Year 1987/70
OXDCs	2.4	3.0	4.7	5.7	6.1	6.3	5.8
OIDCs	3.7	4.7	5.0	5.9	6.5	6.7	3.6
Total Developing Countries	6.1	7.7	10.6	11.6	12.6	13.0	4.6

The table below shows that total developing country oil demand growth was close to 8% per year in the early 1970s. Up until the early 1980's, the OXDCs continued to maintain a demand growth of the same order, whereas OIDCs halved their growth rate after the first round of oil price increases of 1973/74. After the second price round, in 1979/80, OIDCs reduced their oil demand growth rate to only 1% per year. Since the 1985 price decline, OXDCs demand appears to be on a relatively low growth path, while the demand in OIDCS has expanded more vigorously. In the near future, oil demand growth of all developing countries is expected to be of the order of 2.5% per year.

Table 6 **Annual Growth in Oil Demand**

	OXDCs	OIDCs	All Developing Countries
Phase 1 1970-73	7.9%	8.3%	8.1%
Phase 2 1973-78	7.9%	3.7%	5.4%
Phase 3 1978-82	7.2%	0.9%	3.8%
Phase 4 1982-85	2.2%	1.1%	1.7%
Phase 5 1985-87	1.1%	5.3%	3.2%

Source: Secretariat Estimates.

Therefore the OXDC share of total developing country oil demand has increased from 39% in the early 1970's 48% in 1987.

Coal

The CPEs are the largest producers and consumers of coal, led by China which is the world's largest producer and consumer (539.0 Mtoe in 1986). In terms of volume, China overtook the United States in 1983. Its production is almost exclusively used for indigenous consumption, though increased exports are planned. In 1986, the USSR and China together provided more than 40% of total world production. Altogether 46% of primary energy in the CPEs comes from coal while in the Asia-Pacific region, where the market share of coal is next highest, this ratio is 38%. The Middle East by contrast is the region with the smallest market share for coal.

Table 7 **Coal Production 1975-1986**
(Mtoe)

	1979	1985	1986	%Change 1986/85	%Change p.a. 1986/79
Latin America	8.1	16.9	18.1	7.2	7.6
Africa	46.7	94.2	99.1	5.2	7.1
Asia-Pacific	95.0	116.0	121.1	4.4	2.2
CPEs	977.4	1222.9	1270.1	3.9	2.4
Total Non-OECD	1127.2	1450.0	1508.4	4.0	2.7
OECD	640.7	776.3	813.5	4.8	2.2
Total	1767.9	2226.3	2321.9	2.5	2.5

Sources: Energy Balances of OECD Countries, BP Statistical Review of World Energy.

Table 8 **Coal Consumption 1975-1985**
(Mtoe)

	1979	1985	1986	%Change 1986/85	%Change p.a. 1986/79
Latin America	13.3	21.8	22.1	1.2	4.7
Middle East	-	2.2	2.2	−0.5	-
Africa	43.1	65.5	66.2	1.0	4.0
Asia-Pacific	74.0	144.4	150.8	4.4	6.7
CPEs	937.1	1203.0	1235.7	2.7	2.6
Total Non-OECD	1067.5	1436.9	1477.0	2.8	3.0
OECD	624.8	815.5	824.5	1.1	2.6
Total	1692.3	2252.4	2301.5	2.2	2.8

Sources: Energy Balances of OECD Countries, BP Statistical Review of World Energy.

Natural Gas

All non-OECD regions have seen significant increases in both production and consumption of natural gas since 1975. The CPEs had the strongest growth in production between 1985 and 1986 (6.5%) and the strongest growth in consumption, along with Africa (6.7%), with African demand growing from a much smaller base. Indigenous gas supplies are increasingly being used as a substitute for oil, to permit either reduced oil imports or increased oil exports. The large significant decrease in gas production in Africa between 1985 and 1986 reflects in part the reduced output in Algeria as export contracts were being renegotiated.

Table 9 **Natural Gas Production 1975-1986**
(Mtoe)

	1979	1985	1986	%Change 1986/85	%Change p.a. 1986/79
Latin America	42.7	68.2	67.7	−0.8	4.3
Middle East	36.7	46.1	48.0	4.0	2.5
Africa	12.7	44.7	42.8	−4.2	11.7
Asia-Pacific	18.3	70.6	74.3	5.2	13.6
CPEs	318.1	649.6	692.0	6.5	7.3
Total Non-OECD	428.5	879.2	924.8	5.2	7.2
OECD	658.1	637.1	624.7	−2.0	0.5
Total	1086.6	1516.3	1549.5	2.2	3.3

Sources: IEA/OECD Energy Balances; BP Statistical Review of World Energy.

Table 10 **Natural Gas Consumption 1975-1986**
(Mtoe)

	1979	1985	1986	%Change 1986/85	%Change p.a. 1986/79
Latin America	39.2	70.5	71.2	1.1	5.6
Middle East	26.2	46.1	48.0	4.1	5.7
Africa	4.3	26.7	28.5	6.7	18.8
Asia-Pacific	12.2	34.1	35.8	5.0	10.3
CPEs	297.3	578.6	617.3	6.7	6.9
Total Non-OECD	379.2	756.0	800.8	5.9	7.0
OECD	665.0	718.0	701.8	−2.3	0.8
Total	1044.2	1474.0	1502.6	1.9	3.4

Sources: IEA/OECD Energy Balances; BP Statistical Review of World Energy.

Electricity

Electricity generation and consumption grew strongly in the developing countries between 1984 and 1985, led by the Asia-Pacific region (+7.0%). Africa, the Asia-Pacific and the CPEs all rely heavily on fossil fuel-fired generation. Latin America, however, is more dependent on hydro-electricity. The share of nuclear generated electricity as a per cent of total electricity generated is the highest in the CPEs, about 9 %, with Asia-Pacific region, the next biggest, with 7%.

Table 11 **Electricity Generation**
(million kWh)

	1984	1985	% change 1985/84
Latin America	473,663	497,554	5.0
Africa	223,968	228,198	1.9
Asia-Pacific	657,987	704,349	7.0
CPEs	2,408,158	2,504,491	4.0
Total Non-OECD	3,763,776	3,934,592	4.5
OECD	5,692,415	5,902,536	3.7
Total	9,456,191	9,837,128	4.0

Sources: IEA/OECD Energy Balances, United Nations.

Table 12 **Percentage Share of Electricity Generation in 1985**

	Fossil Fuel	Hydro	Nuclear	Geothermal
Latin America	37	62	1	-
Asia-Pacific	69	23	7	1
Africa	77	21	2	-
CPEs	77	14	9	-
OECD	59	20	21	-

Source: IEA/OECD Energy Balances, United Nations.

Thirteen non-OECD countries currently have nuclear generating capacity. The thirteen include the Soviet Union (36 688 MWe), Taiwan (4 884 MWe), South Korea (5 380 MWe), Czechoslovakia (3 160 MWe), South Africa (1 840 MWe), Bulgaria (2 713 MWe), German Democratic Republic (1 702 MWe), Hungary (1 640 MWe),

India (1 464 MWe), Argentina (935 MWe), Yugoslavia (632 MWe), Brazil (626 MWe) and Pakistan (125 MWe). A further six non-OECD countries have plans for nuclear energy before the end of the century.

IV. TRENDS IN ENERGY TRADE

As shown in Table 13, 1292.9 Mmt (an average of 26.27 mbd) of crude oil and products were traded in 1986, up from 1264 Mmt (25.6 mbd) in 1985 .

Table 13 **Oil Trade 1986**
(Mmt)

	Crude Imports	Product Imports	Crude Exports	Product Exports
Latin America	61.7	18.5	126.1	56.7
Middle East	0.9	8.7	391.8	54.8
Africa	22.1	6.3	184.3	19.8
Asia-Pacific	96.8	21.9	50.5	75.0
CPEs*	50.0	5.8	76.9	23.7
OECD	741.0	226.3	57.1	71.4
Destination not known	8.6	22.8	-	-
Total	981.1	311.8	981.1	311.8

* Excludes intra-regional trade. Intra-regional trade included in other regions.

Source: BP Statistical Review of World Energy.

For coal, Table 14 shows the regional trade for coal in 1986 and 1987. Total trade amounted to 337.5 Mmt in 1986, an increase of 1.7 Mmt compared with 1986. World coking trade, representing about 48% of total world trade, actually increased 1.6% to 162.1 Mmt between 1986 and 1987 while steam coal trade decreased 0.5% to 175.4 Mmt.

Table 14 **World Total Coal Trade**
(million metric tons)

Exporters	N. America 1986	N. America 1987	OECD Europe 1986	OECD Europe 1987	Japan 1986	Japan 1987	Latin America 1986	Latin America 1987	Asia* 1986	Asia* 1987	Afr.+M.E. 1986	Afr.+M.E. 1987	CPEs 1986	CPEs 1987	Bal. Item 1986	Bal. Item 1987	World 1986	World 1987
Canada	0.4	0.5	2.4	2.3	17.3	16.6	1.4	1.3	4.1	5.1	-	-	-	-	0.1	0.9	25.9	26.7
United States	13.3	14.3	38.9	30.9	12	9	6.3	6.2	6.9	8.1	2.3	2.1	1.9	1.9	-4.4	-0.3	77.6	72.2
Australia	-	0.2	22.8	26.8	42.2	47	0.8	1.8	22	22.6	2.2	1.8	0.9	1.9	0.5	-0.9	91.9	101.2
Other OECD	-	-	11.7	11.2	0.3	0.3	-	-	-	-	0.1	0.1	0.6	0.6	2.2	-0.1	14.5	13.4
OECD	13.7	15	75.8	71.2	71.8	72.9	8.5	9.3	33	35.8	4.6	4	3.4	4.4	-1.6	0.5	209.9	213.1
Poland	-	-	14.1	14.6	-	-	2.6	2.6	0.3	0.3	0.1	0.1	17.3	12	-	-	34.4	29.6
U.S.S.R.	-	-	4.5	5.4	5.1	6.2	-	-	2.8	3.9	-	-	15.9	15.3	-	-	25.4	27
China	-	-	0.8	1.6	3.6	3.8	-	-	0.3	0.4	-	-	2.9	3.7	-	-	9.9	13.1
Colombia	0.7	0.9	3.5	4.9	0.2	0.1	0.5	1.3	-	-	-	0.4	-	-	-	0.7	5.4	8.6
South Africa	0.9	-	23.3	20.2	8.9	7.3	-	-	11.3	9.8	2.1	3.0	-	-	-	0.4	45.5	40.7
Other Non-OECD	-	0.1	1.7	1.6	0.8	0.8	-	-	1.3	1.3	-	-	1.3	1.6	-	-	5.2	5.1
TOTAL	15.3	16	123.7	119.5	90.4	91.1	11.6	13.2	49	51.5	6.8	7.5	40.8	37	-1.6	1.4	335.7	337.2

* Excludes Asian CPE's and Japan.
Note: Data in the columns N. America to Japan are import statistics. The rows Canada to Japan are import statistics. The rows Canada to OECD (except for above mentioned columns) are export statistics.

Sources: IEA/OECD Coal Statistics and other Secretariat sources, including international sources.

V   THE MAJOR FACTORS AFFECTING ENERGY PRODUCTION AND CONSUMPTION

At the best of times, energy policy is formed and energy projects are undertaken under conditions of great uncertainty. These include economic uncertainties such as the rate of economic growth but also political uncertainties. In the current period, there are a number of factors which energy investors and policy makers must also consider, including:

— *Assessing energy requirements.* Projecting energy demand with the purpose of determining electricity generating capacity requirements, for example, has been more difficult in the past 10-15 years than previously. This is partially due to changes in energy prices which have affected efficiency and fuel switching, and thus the absolute and relative components of the fuel mix. Because of the capital cost of many energy projects, a more complete understanding of energy use and demand is needed.

— *Higher costs of capital investment for new production and transformation facilities.* Exploration and development is often becoming more expensive as the lower cost reserves are exploited and activity moves to develop higher cost reserves.

— *Investment for energy conservation measures becomes more expensive per unit of benefit* after lower cost measures are completed.

— *Increasing concern about the environment* often brings about higher capital and maintenance costs and/or policies to reduce energy demand or change the fuel mix.

— *Uncertain energy prices lead to indecision about further exploration and development, fuel diversification and conservation.*

— *Debt servicing requirements* of many non-OECD countries may affect investments in new energy projects.

PART II

I. EFFECTS OF LOWER OIL PRICES ON ENERGY PROSPECTS FOR NON-OECD COUNTRIES

A. Introduction

Since the major oil price decline which started in early 1986, there has been considerable concern about the effects on future energy production, consumption and trade both for OECD and non-OECD countries. This chapter attempts to examine the short and medium term effects on non-OECD countries: oil importers, exporters and CPEs. Because of the key importance of oil, this chapter will restrict itself to the prospects of oil — in the light of what is likely to happen to other fuels.

B. Developing Countries Oil Production

Developing countries' oil production is concentrated in Iran, Iraq, Kuwait, Saudi Arabia and the United Arab Emirates (UAE) among the Middle East OXDCs; in Mexico, Venezuela, Indonesia, Egypt, Algeria, Libya and Malaysia among the other OXDCs and in India, Brazil and Argentina among oil importers (Argentina is roughly self-sufficient in oil). Oil production outside the Middle East grew by an average 5.0% per year during the second half of the 1970s with Mexico, Egypt,

Indonesia, Libya, Nigeria and Malaysia accounting for much of the increase. Their production growth then slowed to an average 0.1% per year during the first half of the 1980s. During this period, more countries participated in the expansion of non-Middle East developing country production led by Mexico, Brazil, Egypt, Cameroon, Oman, India and Malaysia. Table 15 provides historical production figures for selected developing countries' oil production.

For the future, the rate of growth of oil production is likely to continue to moderate due to factors bearing on production in individual key countries:

— *In Latin America,* Mexico's financial difficulties may limit its ability to finance substantial increases in production capacity, particularly while there is substantial excess production capacity worldwide. Beyond 1987, Mexico is expected to increase production in order to accommodate both an increase in domestic consumption of 0.2-0.3 mbd and to allow for an increase in exports of about 0.3 mbd in order to support renewed economic growth. Venezuela has a capacity of 2.6 mbd, but its production of conventional crude is unlikely to exceed its self-imposed production ceiling of 2.3 mbd. Brazil's production has recently expanded rapidly, and further increases in capacity will result from the ongoing development of the Campos Basin, but mainly in deeper water of over 400 metres requiring further technological advances. Colombia's production is expected to expand still further with developments in the onshore Llanos Basin. Ecuador may be able to continue on an upward path with more exploration successes. Argentina, like Mexico, may have difficulty financing its exploration programme, but is relying more on outside risk investment to help sustain and expand production; it may be difficult for Argentina to sustain recent production levels.

— *In Africa,* Egypt, due to its low reserves/production (R/P) ratio, is unlikely to expand significantly its production capacity from its current 0.9 mbd unless its recent increase in exploration activity provides more successes. Unless new reserves are found, Algeria is likely to concentrate its exports on natural gas and NGLs, while Libya may not need to expand production with its smaller population. Gabon has had recent successes and should be able to exceed 0.2 mbd by 1990. Production in Cameroon and Congo may stabilise at recent levels, or may decline if new developments are not sufficient to offset declining production in mature areas.

Table 15 **Oil Production* in Selected Developing Countries, 1975-1987**

	1975	1980	1983	1984	1985	1986	1987
				(000 b/d)			
Africa							
Algeria**	1020	1230	981	954	950	975	1085
Angola**	159	158	180	209	239	273	342
Cameroon	0	60	123	135	184	186	171
Congo	34	59	105	117	130	129	120
Egypt	297	590	735	862	905	819	932
Gabon**	225	175	155	168	170	160	156
Libya**	1480	1830	1109	1114	1075	1066	1007
Nigeria**	1785	2055	1246	1395	1496	1441	1277
Tunisia	95	100	115	116	115	115	104
Sub-total	5095	6257	4749	5070	5264	5164	5194
Asia-Pacific							
Brunei	181	230	169	157	155	170	145
India	169	193	464	525	599	600	623
Indonesia**	1305	1575	1402	1499	1347	1429	1370
Malaysia	94	276	370	424	401	483	465
Sub-total	1749	2274	2405	2605	2502	2682	2603
Latin America							
Argentina	396	491	488	489	480	459	447
Brazil	172	187	328	471	564	608	596
Chile	21	33	40	40	40	35	30
Colombia	158	125	156	167	182	298	389
Ecuador**	160	205	225	253	283	267	173
Mexico	806	2130	2951	2979	2992	2718	2844
Peru	72	191	177	180	200	180	172
Trinidad	215	215	162	160	170	170	162
Venezuela**	2425	2235	1866	1874	1711	1801	1848
Sub-total	4425	5812	6393	6613	6622	6536	6661
Middle East							
Bahrain	57	48	50	49	49	49	49
Iran**	5385	1480	2471	2183	2258	1856	2280
Iraq**	2260	2645	1000	1192	1407	1703	2103
Kuwait**	1885	1430	913	974	923	1305	1196
Neutral Zone	500	545	396	400	357	344	365
Oman	341	283	391	413	496	555	573
Qatar**	435	460	321	421	316	356	307
Saudi Arabia**	6970	9990	5175	4714	3598	5182	4447
Syria	181	160	169	180	180	199	222
UAE**	1655	1695	1231	1394	1332	1596	1724
Sub-total	19669	18736	12117	11920	10916	13145	13266
Other	64	115	138	136	159	208	183
TOTAL	30982	34174	25803	26343	25464	27734	27908

*   includes NGLs
** OPEC members

Sources: IEA Database; BP Statistical Review of World Energy.

Nigeria is likely to wish to expand production to 2 mbd as market demand develops. Tunisia's production is not expected to increase significantly although resolution of an offshore boundary dispute with Libya could allow a modest increase in production.

— *In South Asia,* India's production is expected to rise to about 0.65 mbd by 1990 through further development. And in the Far East, Malaysia has increased its production to 0.5 mbd, and may be able to maintain capacity at or above this level through increasing offshore developments as older, onshore fields decline. Indonesia could again exceed previous peak production of 1.7 mbd if current exploration efforts are successful, but it may instead try to hold a production level of about 1.5 mbd.

— *In the Middle East,* Oman's production may rise, but is expected to remain below its estimated 0.65 mbd production capacity. Syria's production is assumed able to increase on the basis of recent discoveries. Further, North Yemen started production of about 0.15 mbd in late 1987; this is to reach 0.2 mbd by mid-1988 and possibly up to 0.3-0.4 mbd by 1990. Qatar has about 10 times in the way of natural gas reserves as oil reserves, and may therefore not wish to expand oil production past 0.3 mbd. Apart from the special case of the Neutral Zone, which is likely to be permitted to produce 0.4 - 0.5 mbd, there are 5 other major oil producers in the region (Saudi Arabia, Iran, Iraq, Kuwait, and UAE), with R/P ratios in excess of 60 years. For the near-term, Iran and Iraq have war-related constraints on sales, but past experience suggests that the other 3 countries could expand production to 50-100% above their present levels, as demand increases, i.e. to 6-8 mbd for Saudi Arabia, and 1.5-2.0 mbd for Kuwait and the UAE.

In addition to the impact of cashflow on exploration expenditures, lower oil prices can have potential impacts on production from developed fields and on the pace of development of new fields. For most developing countries, the combination of relatively modest production costs in onshore and shallow offshore fields and the pressing need for oil revenue suggest that the short-term production response to lower oil prices may be slight; moreover there is considerable momentum to development projects under way. In the extreme, some smaller countries with shut-in capacity could actually increase production in response to lower prices in order to maintain revenues.

Current oil industry cash flow has less impact on development expenditures than on exploration. Exploration is generally financed from cashflow whereas development often involves financial institutions which evaluate the project in terms of expected rate of return, not current cash flow. Since development may be less sensitive to current cash flow than exploration, there is likely to be certain momentum to the increase in oil production for the rest of the 1980s based on the developments of already known resources. In the 1990s, the impact of cash flow in reducing exploration is assumed to result in fewer discoveries, and production capacity is assumed to grow slowly at prices below $25 per bbl. Exploration investments will tend to be in regions other than the Middle East, as that is where the spare capacity is least and the R/P ratios are lowest; (except for Libya, Mexico and Venezuela, the R/P ratios in developing countries tend to be in the 10-30 year range). In respect of the Middle East, 1986 production of 13.0 mbd represents little over half (58%) of all-time maximum production of 22.5 mbd in 1977. Current spare capacity in the Middle East is of the order of 6-7 mbd. On the other hand, 1986 production outside the Middle East of 12.1 mbd represented about 80% of current sustainable capacity of around 15.0 mbd.

For many countries, it is not the physical resource base which determines present and future preferred production levels but:

— *Revenue requirements,* which are determined in part by the size of their population (and hence their absorptive capacity for spending). They are also determined in part by desired investment internally on infrastructure and industrial development, and in part by defence requirements and aid policies;

— *Debt servicing requirements;*

— *Overseas investment plans* (such as downstream in oil refining and marketing), or a desired build-up of other foreign assets;

— *the protection of the existence of an adequate-sized oil market* at reasonable selling prices into the future, so that their oil resource base continues to have saleable potential; and

— *social and political factors.*

Matters which concern exploration and production investments outside the Middle East such as costs of production, after tax rates of return, geological prospectivity, and equity rights and obligations tend to assume less importance in the Middle East.

C. Developing Country Oil Consumption

Oil consumption in developing countries is generally expected to increase somewhat more in percentage terms than in OECD countries. This expectation is based on a number of factors including: expected higher rates of economic growth than the OECD average; ongoing industrialisation and urbanisation; and growing transportation requirements. The response to lower oil prices conceals the partly offsetting responses of the two constituent parts - OXDCs and OIDCs, which each account for roughly half of total developing countries oil consumption:

— OXDCs have suffered a loss of export revenues. This has, in many cases, necessitated contraction of their domestic economies - including energy and oil consumption. For some heavily indebted OXDCs debt service has become extremely difficult. In the longer term, increasing world oil demand in response to lower prices should improve the prospects for those OXDCs which can increase their oil production levels.

— OIDCs have benefited directly from lower oil import expenditures. At the same time, their own exports have increased significantly and thus improved their trade and current balance. In the longer term, however, lower oil prices can negatively affect the rate of development of indigenous energy resources in OIDCs, and encourage greater oil consumption, and thus continued or increased reliance on imported oil. Certain OIDCs had benefited from aid flows from OXDCs, which have been reduced significantly because of decreased revenue.

In 1986 oil demand in developing countries was estimated at approximately 12.6 mbd, roughly 21% of world oil consumption. For a number of reasons, their oil use may increase more rapidly during the next 15 years than oil use in OECD countries or CPEs. These include:

— many developing countries, particularly many of the OXDCs, have very low domestic petroleum product prices which encourage rapid consumption growth;

— some developing countries are experiencing rapid population growth, increasing urbanisation, or both. These can contribute to rapid growth in petroleum consumption;

— there is decreasing availability of traditional fuels in some regions;

— rising per capita incomes, particularly in the newly industrialised countries, can lead to increasing transportation fuel consumption;

— expanding road systems will lead to increased consumption of petroleum products by commercial vehicles and private automobiles;

— increasing industrialisation and expanding electricity distribution, may lead to increased oil use, depending on the price of oil and the availability and price of alternative fuels.

In general, light product demand growth rates are high in keeping with population growth, increasing transportation requirements and urbanisation. Heavy fuel oil consumption growth is expected to slow in many OXDCs as gas utilisation increases. However, among the OXDCs, lower oil prices are associated with less rapid economic growth. While this results in lower energy consumption in industry, it also reduces funds available for investment in alternative fuels, particularly natural gas infrastructure. These offsetting influences result in a very limited response in fuel oil demand between the scenarios considered for the OXDCs as a group. For OIDCs, considerable progress has been made in substitution since 1979 but some substitution potential will remain in the 1990s. Thereafter, heavy fuel oil use is expected to exceed economic growth if oil prices remain low due to expected growth in electricity consumption at rates exceeding the overall economic growth rate in developing countries. However, rising oil prices will in due course prevent rapid growth in heavy fuel oil consumption in the OIDCs.

For all products except heavy fuel oil, historical consumption growth was compared to measures of economic growth. In general, between 1979 and 1984, oil consumption (excluding heavy fuel oil) grew more rapidly than real GDP for the oil exporters and less rapidly for the oil importers. In part, this was due to the low domestic petroleum product prices in many OXDCs, compared to higher prices in OIDCs.

Table 16 below compares historic growth in light petroleum product demand and estimated GDP growth for several groups of developing countries and centrally planned economies. In general, if oil prices are high, continued conservation and fuel substitution will reduce the rate of growth of light product consumption relative to GDP growth.

For lighter products OXDCs are expected to reduce their very high historical rates of consumption growth to near the economic growth rate by 1990. Low domestic prices are expected to prevent a further slowing of consumption/GDP growth ratios in OXDCs. OIDCs are expected to experience a sharp increase in light product consumption growth for a number of reasons including: more rapid economic growth; the substantial fall in oil import prices since late 1985; increasing transportation requirements associated with rising living standards. However, rising oil prices in the 1990s would reduce the rate of light product consumption growth.

Table 16   Comparison of Growth in Consumption of Light Petroleum
Products and GDP Growth, 1980-1984
(%)

	Average Annual Growth in Light Product Demand (1)	Average Annual Growth in GDP (2)	Differences (1) - (2)
OECD	−2.1	1.9	−4.0
Centrally Planned Economies			
USSR	2.5	2.1	0.4
China	0.8	5.2	−4.4
East Europe	−1.3	0.9	−2.2
Oil Importing Developing Countries			
South Asia	5.7	5.7	0
East Asia	4.3	5.4	−1.1
Brazil	0.7	1.5	−0.8
Other Latin America	−0.1	0.4	−0.5
Africa & Middle East	3.4	2.8	0.6
Oil Exporting Developing Countries			
Asia & Pacific	6.3	6.2	0.1
Mexico	6.9	2.7	4.2
Other Latin America	0.9	−0.3	1.2
Middle East	7.5	−1.6	9.1
North Africa	9.9	2.0	7.9
Africa	6.6	−2.9	9.5

Source: Secretariat estimates.

D.   CPE Oil Consumption and Net Exports

CPE oil consumption is expected to increase at very modest rates during the medium term due mainly to substitution of natural gas for heavy fuel oil in the Soviet Union and Eastern Europe. The Soviet Union is thought to be capable, with adequate investment, of reducing its heavy fuel oil consumption from 3.0 mbd in 1984 to under 1.7 mbd during the next 10 years. Eastern Europe is expected to proceed with gas, coal and nuclear programmes under way leading to a 20% reduction in heavy fuel oil use before its consumption begins to grow apace with economic growth in the mid-1990s. China's economy is essentially coal based but some oil substitution possibilities exist.

Light product consumption growth is assumed to slow in the Soviet Union in order to conserve oil for export purposes. In Eastern Europe, decreasing availability of bargain-priced Soviet oil is assumed to lead to strong conservation measures and almost no growth in consumption. In China, oil consumption is assumed to increase at about half the rate of economic growth. This is a low rate which assumes restraint of growth of transport sector demand for oil in order to maintain or even increase oil exports.

Table 17 indicates production, consumption and net export levels for CPEs for 1985 to 1987 and forecast for 1990. Oil production to 1990 is assumed to increase from 2.5 to around 2.85 mbd in China and to stagnate in Eastern Europe at 0.4 mbd. Table 18 shows the historical development of Soviet oil production. For the Soviet Union, the current heavy investment in oil production is assumed to give way to a more balanced investment programme in the face of lower oil prices. Accordingly, oil production is assumed to decline throughout the medium term if the price of oil remains low. If there is an increase in oil prices in the 1990s it is assumed they will cause the Soviet Union to redirect its investment toward the oil sector and to maintain production at near present day levels throughout the first half of the 1990s.

For the Soviet Union, the investment costs of maintaining current levels of oil production compete for limited funds with alternative investments in modernisation of the economy. The Soviet Union is assumed to limit its investment in maintaining oil production levels, particularly its hard currency investments in production equipment, in response to lower oil export earnings having regard to the rising marginal cost of oil production. This reduction is expected both as a response to reduced hard currency earnings and to the decreased convertible currency returns to oil production. It is possible that the Soviet Union, with limited options in effect, would attempt to maintain hard currency earnings from oil, irrespective of domestic costs.

Table 17 **CPE Oil Production, Consumption and Net Exports**
(mbd)

	1985	1986	1987	1990
USSR				
Production	12.22	12.55	12.73	12.85
Consumption	9.18	9.32	9.39	9.25
Net Exports	3.04	3.23	3.34	3.60
China				
Production	2.50	2.61	2.70	2.85
Consumption	1.76	1.94	2.05	2.40
Net Exports	0.74	0.67	0.65	0.45
East Europe				
Production	0.40	0.40	0.40	0.40
Consumption	2.10	2.10	2.10	2.15
Net Exports	−1.70	−1.70	−1.70	−1.75
Total CPEs (except Cuba and Asian CPEs)				
Production	15.12	15.56	15.83	16.10
Consumption	13.07	13.36	13.54	13.80
Net Exports	2.05	2.20	2.29	2.30
Net Exports to Other CPEs and Statistical Difference	−0.2	−0.3	−0.2	−0.3
Net CPE Exports	1.8	1.9	2.1	2.0

Table 18 **Soviet Oil Production, 1960-1987**

	1960	1965	1970	1975	1980	1986	1987
Production (mbd)	2.9	4.9	7.1	9.9	12.14	12.55	12.73
Average Growth (%)	n.a.	10.4	7.6	6.8	−0.2	3.4	1.4

For China, current developments are expected to take production to near 3.0 mbd by 1990 even in a low price scenario. Beyond 1990, the impacts of reduced oil company cash flow on exploration in China are expected to lead to a decline in production in the lower price scenario.

What has been apparent is that, notwithstanding rising oil prices in real terms and substitution for heavy fuel oil during the 1980s, developing countries oil consumption has continued to increase strongly, and is likely to continue to do so under the conditions envisaged. Oil production is also likely to expand, particularly up to 1990, the rate thereafter depending on the level of the oil price.

For the CPEs, while oil demand growth is likely to be very low, only China is likely to be able to increase oil production significantly. As a consequence, net CPE oil exports are likely to decline or even disappear during the 1990's, with Eastern European imports exceeding the aggregate of Soviet and Chinese exports.

E. Financial Aspects of Lower Oil Prices

Among non-OECD countries, the lower oil prices have mainly had a negative financial impact on OXDCs. As seen in Part I, net oil export earnings fell $78.4 billion, or over 46%, between 1985 and 1986. The impact of this loss in export earnings on the domestic economies of the individual exporting countries depends on the importance of oil earnings relative to other exports and to total GDP.

A number of developing countries, particularly Algeria, Brunei, Indonesia, and Malaysia export substantial amounts of natural gas, and the Soviet Union is the world's largest natural gas exporter. The revenues from natural gas exports declined in 1986 in response to falling world oil prices, though with a delay up to three to five months. Prices in natural gas export contracts, whether for pipeline gas or for liquified natural gas, are increasingly related to the price of competing fuels.

OXDCs have generally responded to the loss in oil revenues by reducing imports and contracting the domestic economy. The economic contraction can be moderated, in some cases, by alternative policies. For example, an oil exporting country can spend from accumulated financial assets. Saudi Arabia, for example, has run substantial current account deficits in recent years, greatly reducing the assets it had accumulated since 1974.

Other countries without such assets have had to increase their borrowings and reschedule their debt repayments. Mexico, for example, has recently negotiated an agreement with the International Monetary Fund for substantial new loans. The agreement provides for an increase in the lending if the oil price falls further.

Countries can reduce both official and private transfers. The recent contraction of the construction industry in the Arabian Gulf has greatly reduced outflows of private remittances by foreign workers. This has had an adverse effect on Egypt, Pakistan, and India, which rely on worker remittances as major sources of foreign earnings. Official development aid by OPEC member countries sharply declined from its 1980 peak of $9.2 billion to $3.7 billion in 1985 (at 1985 prices and exchange rates), mainly in response to falling oil incomes. Aid increase to $4 billion in 1986.

The Soviet Union has relied on oil and gas exports for about three-fifths of its convertible currency earnings. However, the Soviet Union has a large and diverse economy compared to most OXDCs, and may have some policy options in addition to reducing imports and increasing borrowings. For example, the Soviet Union also exports gold, natural gas and arms, although natural gas prices have declined along with oil prices.

Oil importing countries directly benefit from reduced oil import costs. They also benefit from increased exports as the result of stimulated economic activity in developed countries. In addition, indebted OIDCs could benefit from lower interest payments on new borrowings and outstanding variable rate loans, if interest rates decline in response to lower oil prices.

However, there are factors which can partly offset the benefits of lower oil prices to OIDCs. These derive from efforts by some OXDCs to cope with lower oil revenues and include: reduced merchandise exports to OXDCs, reduced worker remittances from OXDCs, and reduced development aid from OXDCs.

F. **Other Aspects**

The decline in oil prices has potential to create other problems for non-OECD countries. Official development aid from OPEC member countries has declined sharply from its 1980 peak as oil revenues have declined. This adds to the economic difficulties of the principal receiver nations of this aid, mainly Arab countries (e.g. Syria, Jordan, Sudan, North Yemen).

The decline in economic activity in Gulf countries has reduced worker remittances, as noted above. In addition, it has led to the departure of many guest workers. These returning workers may add to unemployment problems in their home countries (e.g. Egypt, Pakistan, India). Or, as in the case of Palestinians, departing guest workers may result in an increase in the regional tensions, as well as reduced cash flows into the West Bank and refugee areas.

II. ENERGY DEVELOPMENTS IN THE SOVIET UNION AND CHINA

A. Introduction

China and the Soviet Union represent two of the world's largest and important energy producers and consumers. Comparatively little has been known about their energy sectors, even though, for example, the Soviet Union has been a major exporter of both oil and natural gas for many years.

The IEA Secretariat has been following developments in these two countries. The following sections are short summaries of the major recent developments.

B. Soviet Union

Since 1976 the Soviet Union has been the world's largest oil (crude and NGLs) producer, and since 1983 the world's largest producer of natural gas. However, while its gas production has continued to grow at 6-10%

per year during the 1980s (averaging 7.6% per year), oil production has only grown at an annual rate of 0.5% in the 1980s. Table 19 shows the trends in oil and gas production of the world's major producers.

Table 19 **Oil and Natural Gas Production**

	1960	1970	1980	1987
Oil (mbd)				
Soviet Union	3.0	7.1	12.2	12.7
Saudi Arabia	1.2	3.6	10.0	4.4
United States	8.0	11.3	10.2	9.9

				1986
Natural gas* (bcm)				
Soviet Union	51	213	448	703
United States	378	634	581	483

* excluding reinjected gas.

Sources: IEA data base; BP Statistical Review of World Energy, Cedigaz.

The Soviet Union's oil production depends significantly on the supergiant Samotlor field in the Tyumen Oblast basin, which was discovered in 1965, and started producing in 1969. Samotlor produces about one-fifth of current Soviet oil production.

Official estimates of the Soviet Union's proven recoverable oil reserves are not published. It seems likely that these reserves are of the order of 60 billion barrels - about 7% of world oil reserves - which would give a R/P ratio in the Soviet Union of about 13 years, slightly above that of IEA oil producing countries.

In the absence of any further huge oil finds since Samotlor, apart from the difficult Tengiz field, and to reduce the very large (approaching 40%) share of Soviet investment being devoted to the energy sector, it seems inevitable that Soviet oil production will drift back to about 12 mbd in the 1990s. Unless more success than hitherto is achieved with reducing oil consumption, the Soviet Union will want to attempt to hold oil production at 12 mbd in order to protect export earnings as much as possible.

The Soviet Union has relied on energy exports (mostly oil) for about half of its total export earnings, and for as much as three-quarters of its hard currency earnings from sales to OECD countries (see Table 20). Its oil and gas exports are split roughly in half between Eastern and Western Europe.

Table 20 **Energy Share in Soviet Export Earnings from Major Regions**
(%)

Exports	1970	1975	1980	1982	1984	1985	1986
To East Bloc	14.3	24.7	38.3	48.4	49.5	45.9	47.0
To OECD and Developing Countries	18.2	39.7	55.4	57.5	58.7	60.1	44.4
To OECD	55.8	58.6	70.0	79.7	78.5	75.9	64.3
To World	15.7	30.8	46.2	52.7	53.5	51.4	46.1

Sources: Narkhoz and Wharton, CPE Databank.

Natural gas production has continued to grow as several giant gas fields in North-West Siberia have been exploited - first Urengoi (estimated reserves of 7000 bcm), then Yamburg (estimated reserves of 4500 bcm), and next Yamal. The Soviet Union has about 38,500 bcm of proven recoverable gas reserves (which are the world's largest and amount to about 39% of the world total) - far ahead of second-placed Iran which has 13,900 bcm.

While almost 30% of Soviet oil production is exported, just over 11% of Soviet gas production is exported. As annual Soviet gas production is expected to expand to 1,000 bcm by the mid-1990s, there will be scope for increased Soviet gas exports, particularly to Eastern Europe through the Progress pipeline after 1989.

Hydroelectricity production growth has slowed during the 1980s, from 4.0% per year during the 1970's to around 2.7% per year in the 1980s. Hydroelectricity met about 13% of Soviet electricity needs in 1986. The emphasis has switched to nuclear electricity production, but the ambitious programme has consistently fallen behind plan targets because of construction delays. Then in April 1986, an explosion in Unit 4 of the 4×1000 MW Chernobyl station occurred, further delaying the programme. In 1986, nuclear met about 10% of Soviet electricity needs. Table 21 and Figure 2 show the evolution of the Soviet Union's electricity production.

Table 21 **Soviet Electricity Production**
(Twh)

	1970	**1975**	**1980**	**1985**	**1986**
Hydro	124	126	184	215	216
Nuclear	4	40	73	167	161
Thermal	613	873	1037	1162	1222
Total	741	1039	1294	1544	1664

Sources: Various Soviet publications.

Figure 2 **Soviet Electricity Production**
1970-1986

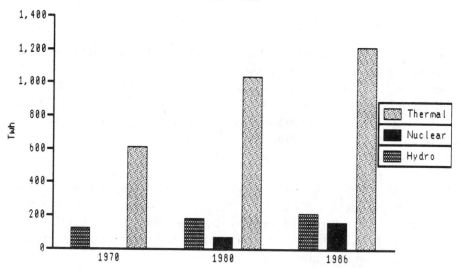

Source: Various Soviet Publications.

Coal production stagnated, fluctuating around 715 Mmt between 1976 and 1984, probably as a result of a low investment priority for coal. Coal quality has been another major problem, which has meant that some coal-fired power stations have had to be switched wholly to gas, or to burning coal spiked with oil. In calorific value terms current coal production is 7% below 1980 levels.

With coal contributing very little to increasing needs for fuels for thermal electricity and heat generation during the 1980s, more than 46% of the expansion in gas supplies has had to be assigned to this sector.

The principal area for oil substitution is heavy fuel oil use, an area where OECD countries have more than halved consumption since 1978. The total reduction to date in heavy fuel oil use in the Soviet Union appears to be about 17 Mmt only from a peak of 150 Mmt over 1980-82.

The lack of success with oil substitution appears to arise because the energy intensity of the Soviet economy has increased by 5% since 1970 when measured by the traditional relationship of total primary energy requirements to GDP - while that in IEA countries has fallen by over 20% (see Table 22). Comparing total primary energy requirements with the preferred Soviet indicator of economic growth of net material product (NMP) shows a reduction of 16% over 1970-86. Progress has been made with reducing oil intensity since 1980, but at only half the rate being achieved in IEA countries.

Table 22 **Energy Demand and Intensity**

	1970	**1975**	**1980**	**1986**
Apparent total primary energy requirements (TPER) (mtoe)	797.9	977.7	1162.0	1318.3
GDP (in 1980 $bn)	811.5	973.8	1112.6	1279.8
Energy intensity TPER/GDP	0.983	1.004	1.044	1.030
NMP (in 1983 Roubles bn)	313.1	392.4	491.8	611.8
Energy intensity TPER/NMP	2.55	2.49	2.36	2.15
Oil requirements (mtoe)	261.8	368.0	448.7	451.0
Oil intensity Oil/GDP	0.323	0.378	0.403	0.352

Source: Wharton CPE Databank.

The relative lack of success with reducing energy and oil use in the Soviet energy economy makes it in the late 1980s, more akin to the IEA energy economies of the early 1970's (see Table 23).

Table 23 **Energy and Oil Intensities**

	1973		1986	
	Energy/ GDP	Oil/ GDP	Energy/ GDP	Oil/ GDP
Canada	0.88	0.39	0.76	0.23
United States	0.76	0.34	0.57	0.24
IEA Pacific	0.42	0.30	0.31	0.16
IEA Europe	0.40	0.23	0.34	0.15
IEA Total	0.56	0.29	0.44	0.19
Soviet Union	0.99	0.35	1.03	0.35

Sources: Secretariat estimates, Wharton CPE Databank.

This lack of progress is due in part to the unchanged weight of heavy industry in the total economy and in part to the absence of a meaningful internal pricing system, which properly reflects costs, and which would persuade energy users to take appropriate economic actions.

C. China

The energy scene in China has been, and continues to be, dominated by coal, which meets about 75% of total commercial primary energy requirements (TCPER)*. While coal's share of TCPER did drop to around 70% of TCPER in the second half of the 1970's, the rising value of oil, and the comparative production potentials for coal and oil, led China to back substitute coal for oil in industrial and power station boilers.

Raw coal production in China has grown by an average of 5.4% annually since 1966, and as shown in Table 24 and Figure 3, has since 1985 been greater than both the Soviet Union and the United States. Measured in

---

*   This excludes non-traded energy.

terms of the energy equivalence (million tonnes of coal equivalent (Mtce), however, the United States still remains the world's foremost producer of coal-based energy, as shown by the last column of Table 24.

Figure 3 **Raw Coal Production**
1960-1987

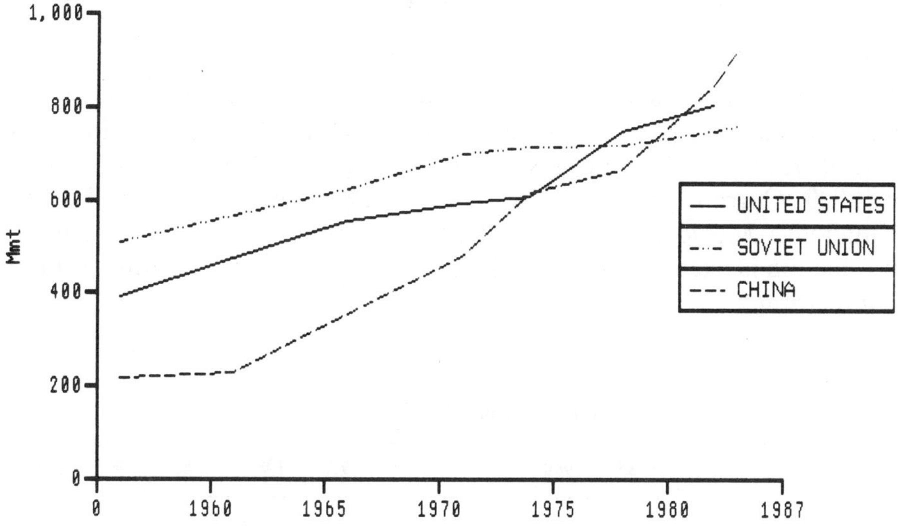

Table 24 **Raw Coal Production**
(Mmt)

	1960	1965	1970	1975	1978	1982	1986	1987	1986 Mtce
China	220	232	354	482	618	666	850	920	630
Soviet Union	510	n.a.	624	701	716	718	751	760	455
U.S.	394	478	556	594	608	750	806	n.a.	669

Source: IEA Database.

Several researchers expect China's coal production to grow to 1200-1400 mtce by the year 2000. This would give a growing exportable surplus of coal, much greater than China's coal exports of 10 Mmt in 1986.

Along with China's substantially growing coal production have come both railway transport bottlenecks and environmental problems. The

— 41 —

former may be rectified by mine-mouth power stations, coastal shipping of coal, improved railway systems, and increased investments in dressing (washing, etc.) to reduce the quantity requiring to be transported. The latter will also be improved by increased dressing (as a better quality product will be burnt), and investments in technology to reduce emissions by coal consumers. Households consumed 23% of coal in 1982 (148 Mmt), and better equipment in that sector, in particular, tends to lead to the cleaner use of coal.

Oil production has also expanded from very low levels in the mid-1950s, and a significant rate of growth of 4.9% annually has been maintained since 1975. Official targets have been to reach 3 mbd by 1990, and 4 mbd by 2000. Current production is about on a par with that of the United Kingdom, Mexico, Iran and Iraq — countries whose individual production is exceeded by only Saudi Arabia, the Soviet Union, and the United States.

Table 25 **China's Oil Production**

	1955	1960	1965	1970	1975	1980	1982	1984	1986
Mmt	1.0	5.2	11.3	30.6	77.1	106.0	102.1	114.7	130.6
Mbd	0.02	0.10	0.23	0.61	1.54	2.12	2.04	2.29	2.61

Source: Statistical Yearbook of China.

Demand for oil peaked in 1978 at about 1.84 mbd, after which it fell back to a low point of about 1.65 mbd in 1982, due almost entirely to the reduced use of crude for direct burning in boilers. Between 1975 and 1985, the demand for transport fuels has grown at a rate of 4.5% annually (from 21.98 Mmt to 34.17 Mmt). There is still the prospect that heavy fuel oil demand in China, which has fluctuated around 27 Mmt (0.5 mbd) since 1976 will be reduced by further substitution by coal. Total oil demand only surpassed the 1978 level in 1986, reaching 1.9 mbd.

Natural gas has not been a major fuel in China, meeting only 2% of primary energy needs, compared with 15-20% in several OECD countries such as Austria, Belgium, Germany and Italy. There is some scope for expanded gas use (largely because of the reserves offshore

Hainan Island) as the R/P ratio is 42 years. There is also the good possibility of substantial further discoveries, as the U.S. Geological Survey estimates that only 14% of China's total gas endowment has been discovered.

China has the world's largest hydropower potential, of which over half (378 GW) is considered exploitable. Of this, only 30 GW has so far been developed. Most of the resources are in the south-west region - rather remote (2000 - 3000 km) from industrial centres in the north-east or in the South Central coastal provinces. Although the rugged terrain hinders development, China plans to complete 10 GW in the 1986-1990 period, and a further 29 GW by 2000.

While no nuclear plants are currently operating, a 300 MW Chinese designed plant at Qinshan is due to come on-stream in 1989, and the 2 × 900 MW Daya Bay project, to supply Hong Kong in part, and being built using French and British technology and plant, is scheduled to start operation in 1992.

Table 26 and Figure 4 show the proportion of electricity production by hydro and thermal means (TWh):

Table 26 **Proportion of China's Electricity Production**

	1980	1983	1985	1987
hydro	58	86	92	97.0
coal-thermal	179	208	{318.3	{399.0
oil-thermal	63	57		
Total	301	351	410.7	496.0
% share hydro	19	25	22	20
% share coal	60	59	{78	{80
% share oil	21	16		

Source: IEA Secretariat Estimates.

Oil represents about 90% of China's energy exports (by energy content), with the remainder being coal. It seems likely that oil demand will rise slightly faster than oil production, resulting in oil exports declining slowly from their 1985 peak. On the other hand, coal exports are expected to rise considerably from a low base, but into a very competitive international market. Coal exports have represented around

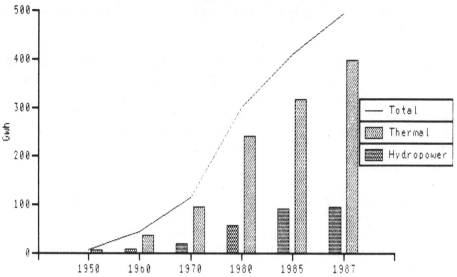

Figure 4 **Electricity Production in China**
1950-1987

Source: Secretariat estimates.

1% of total production, whereas net oil exports amounted to 27% of production in 1986. As a result of lower oil prices in 1986, energy's share of export earnings fell from 20-26% in the first half of the 1980's to 12% (see Table 27).

Table 27 **Net Energy Exports**
(Mmt)

	1975	1980	1982	1984	1985	1986
Oil	11.9	17.0	19.7	27.7	37.2	35.5
Coal and coke (5000 kcal/kg)	3.4	6.6	6.8	7.2	7.7	10
Energy export earnings fob(US$bn)	n.a.	18.5	21.1	23.9	25.1	28.4
Total export earnings(US$bn)	n.a.	4.8	5.0	5.1	5.1	3.3
% of energy in export earnings	n.a.	26	24	21	20	12

Sources: China Statistical Yearbook; China Energy Research Institute;
Ministry of Foreign Economic Relations and Trade;
Wharton CPE Databank.

China's success with substituting coal for oil in TCPER was mentioned above. Table 28 shows the trends, with oil's share of TCPER diminishing from a peak of nearly 23% to 17%.

Table 28 **Shares of TCPER**
(%)

	1965	1978	1982	1985
Coal	86.4	70.7	74.0	75.9
Oil	10.3	22.7	18.7	17.1
Natural gas	0.6	3.2	2.5	2.2
Hydroelectricity	2.7	3.4	4.8	4.8

Sources: Statistical Yearbook of China; Bureau of Energy, State Economic Commission.

As shown in Table 29, considerable progress has been made with reducing energy intensity in China. Energy intensity has fallen by over 30% since 1978. It is still much higher than in IEA countries, where the average energy intensity is 0.44, and that of North America is 0.59. However, oil intensity (oil/GDP) has been halved since 1978: at its present level of 0.20, it is almost the same as the IEA average of 0.19.

Table 29 **Energy Intensity Factors**

	1975	1978	1980	1983	1985
Oil requirements	65.2	91.9	88.9	86.3	87.7
TCPER	318.0	400.0	421.9	462.3	535.0
GDP (1980 US$bn)	172.1	212.8	283.2	335.0	428.7
TCPER/GDP	1.85	1.89	1.49	1.38	1.25
Oil/GDP	0.38	0.43	0.31	0.26	0.20

Sources: Wharton CPE Databank; Bureau of Energy, State Economic Commission; IEA Database.

China has started the process of reforming energy prices with a two-tier price system in force for some forms of energy; customers wanting more energy than the suppliers' quota have to negotiate a price for the excess amount. It is proposed to increase the proportion of energy sold at prices reflecting real costs. Energy price reform is seen as the key to promoting rational energy investments, the efficient use of energy, and the balanced development of energy-intensive industries.

III. ENERGY CONSERVATION POLICIES AND PRACTICES IN SELECTED NON-OECD COUNTRIES

A. Introduction

There is no single reason why countries either have undertaken or should undertake activities to improve energy efficiency. Economic and social structures vary enormously. Energy requirements depend not only on the economic and industrial structure but also on climate, level of modernisation, energy pricing structures, population and size of country to name but a few. To compare countries, the IEA uses the intensity indicator of total primary energy requirements divided by gross domestic product (TPER/GDP) as a "rough" indicator for energy conservation, even though it includes structural change and fuel switching. Table 30 shows, there is a marked difference in the energy used per unit of economic output because of the various reasons listed above — but there is also a difference in how the intensity has changed over time. It is the rate of change that is important in giving an approximation of efficiency improvements. As the table shows, the rate of change varies considerably, although there has been an increase in intensity in most of the OXDCs selected. Most of the OIDCs showed a decrease in energy intensity, although to varying degrees of success.

An important question is how much energy is required to do a certain task and can that amount of energy be reduced in an economic manner. In other words, is there economic potential for energy efficiency improvements. Only detailed country and sector specific analysis can provide that answer. However, some analysis done by individual countries and organisations such as the World Bank do give some indication that there is considerable potential in all end-use sectors and the transformation sector.

Energy conservation activities are very disaggregated. Inevitably decisions on investment and other energy conservation actions rest with industries and consumers. Government policies do, however, provide a framework to encourage such actions. Success requires a number of different approaches and the categories discussed below show the types of actions that lead to improvements in energy efficiency. Considerable conservation activity has taken place in non-OECD countries, particularly since the second major oil price increase in 1979. There is an awareness of the importance of improving energy efficiency — whether it be to improve the environment, solve the fuelwood problem, or reduce expensive energy imports.

Table 30 **Energy Intensity in Selected Countries**
(TPER/GDP)

	1979	1986	% change p.a.
OIDCs			
Argentina	0.28 (1980)	0.33 (1984)	+ 3.3
Brazil	0.54	0.58 (1984)	+ 1.4
Chile	0.40	0.41 (1985)	+ 0.8
India	0.67	0.67 (1984)	-
Korea	0.68	0.65 (1984)	− 0.9
Pakistan	0.91	0.85 (1984)	− 1.4
Singapore	0.66	0.58 (1984)	− 2.6
Thailand	0.73	0.65 (1984)	− 2.3
Tunisia	0.42	0.43 (1983)	+ 0.6
OXDCs			
Algeria	0.35	0.39 (1982)	+ 3.7
Colombia	0.68	0.70 (1984)	+ 0.6
Malaysia	0.53	0.51 (1985)	− 0.6
Mexico	0.53 (1980)	0.55 (1984)	+ 0.9
Saudi Arabia	0.30 (1980)	0.41 (1983)	+ 11.0
Venezuela	0.57 (1980)	0.71 (1984)	+ 5.6
CPEs			
China	1.89	1.25 (1985)	− 5.7
Poland	0.96	0.95	-
USSR	1.04	1.04	-
IEA regions			
North America	0.72	0.58	− 3.0
Pacific	0.38	0.31	− 2.9
Europe	0.37	0.33	− 1.6
Total	0.52	0.43	− 2.7

Sources: Secretariat estimates.

Recently the IEA undertook its own comprehensive analysis of energy conservation policies and practices in its member countries*. The study was completed in early 1987 and was timely because it assessed the role of energy conservation in the context of general energy policy after more

---

*   IEA, *Energy Conservation in IEA Countries,* OECD, Paris, 1987 than ten years of concerted efforts by consumers and governments. It was also timely because many consumer energy prices were falling, after years of continuing increase, and it was unknown whether there remained significant opportunities for efficiency improvement. While the study is limited to IEA countries, the framework and findings may be of use to non-OECD countries who are either developing or assessing their own policies.

The study gave 5 main reasons why energy conservation continues to be important for long-term economic well-being and security:

— energy conservation will extend the availability of energy resources that are depletable;

— there could be a return to tightening energy markets before the end of the century and energy conservation will both delay and lessen its impact;

— energy conservation reduces the environmental consequences of energy production and use in a way which is consistent with energy policy objectives;

— investment in energy conservation at the margin can in many cases provides a better return than investment in energy supply;

— investment in energy conservation can often be undertaken in small increments and is therefore flexible at a time when the energy outlook is uncertain.

Added to these, for many non-OECD countries, particulary those dependent on energy imports, is that improved efficiency can positively affect balance-of-payments or debt problems. Improved efficiency also allows energy exporters to use less and export more. Reducing energy costs can improve industrial competitiveness. Improving efficiency can also reduce the need for non-commercial energy sources such as wood, especially where there may be deforestation problems. And not least of all, as economies are modernising, it is advantageous to build "efficiency" in at the beginning.

Even after the efficiency improvements already made (energy intensity for the whole IEA was reduced 20% between 1973 and 1985), there is still considerable potential for more economically-viable conservation in the IEA countries. Some of this will be achieved in the normal course of events but the study emphasizes that not all will be achieved. The key problem for IEA Member governments addressed in the study was how to achieve more of this economic potential for energy conservation. The achievement of greater energy efficiency is difficult because energy conservation activities are quite disaggregated. Decisions about energy investment and use are made by millions of individual energy consumers. Numerous types of businesses and institutions are involved in promoting energy conservation: companies providing energy conservation services and equipment, public utilities, private, non-profit groups, governments and other concerns for whom conservation activities are only a small part of their total activities.

However, there are many barriers which prevent the economic potential from being fully realized. These barriers include lack of information and technical skills, imbalances between investment in energy conservation and energy supply, invisibility of energy consumption and conservation, credibility of new conservation products, access to capital and so on*.

Success in energy conservation requires a number of different approaches in order to reach and motivate a broad range of interests, for many of which conservation is a secondary activity. As described in the IEA study, the major elements of an effective government conservation policy include:

— energy price, taxation and other policies which give the right economic signals to consumers;

— programmes to reduce or counterbalance limitations which prevent the market from working effectively with respect to energy conservation;

— research, development and demonstration to develop and apply more energy-efficient technologies; research in the social sciences to help provide a better understanding of factors influencing consumer behaviour and to improve the effectiveness of conservation programmes;

— a strong lead by governments in effecting conservation in their own activities;

— effective involvement of the various organisations which work on energy conservation. These include utilities, companies which provide energy conservation equipment and services, voluntary groups and governments.

---

* For a complete review see *Energy Conservation in IEA Countries*.

The categories in the IEA study provide a useful framework to look at the policies developed by non-OECD countries.

Energy pricing and taxation policies

Economic energy pricing is an essential prerequisite for effective energy conservation. There is general agreement among IEA countries that where world markets exist consumer prices should reflect the world market price; in other cases consumer prices should normally reflect the long-term cost of maintaining the supply of the fuel concerned; and proper weight should be given to energy policy objectives in tax policies.

Many countries regulate consumer energy prices. As stated in a recent study by the Asian Development Bank* "In all countries, governments take account of political and social considerations, as well as costs, in setting energy prices. Fixing energy prices below cost is common, and discriminating among fuels and categories of energy users for policy reasons is universal."** Of the 9 countries studied in their report, most notably the Philippines, Korea and Taiwan "used energy prices to encourage energy conservation . . ."** These countries were in the middle to high income categories. Many of the nine countries passed through the energy price increases more after the second major oil price increase in 1979 than after the first in 1973. This phenomenon is confirmed in a separate study*** which compared 8 developing countries and 5 developed countries.

Government Conservation Programmes

Since the first major oil price increase of 1973-74, IEA Member governments have seen a gradual evolution in their policies directed specifically at improving energy efficiency. The first government

---

*   Asian Development Bank, *Energy Policy Experience of Asian Countries*, Manila, 1987. The study analyzed Bangladesh, India, Korea, Nepal, Pakistan, Philippines, Sri Lanka, Taiwan and Thailand.

**   *Ibid.*, p. 91.

***   Gerald Leach et al., *Energy and Growth: A Comparison of 13 Industrial and Developing Countries*, London, 1986.

programmes to enhance energy conservation were broad efforts to educate consumers about simple techniques to reduce energy use and to motivate them to use such techniques. These information campaigns were sometimes supplemented by large financial incentive programmes designed to encourage conservation investments. These actions were initially taken when oil prices were escalating and fears about security were high. Often these first efforts were not as concerned with efficiency improvements as they were with reducing energy demand for cost and security reasons. Three main types of conservation programmes have been used:

— information programmes including audits, labelling, publicity, training and education;

— financial incentives in the form of grants, tax incentives and soft loans; and

— regulations and standards.

Information Programmes

Information programmes are "the cornerstone of all Members' energy conservation programmes."* Information programmes motivate energy users but they are also used to overcome many of the market barriers which impede efficiency improvements. They can be aimed at either the public at large or at specific groups of consumers or providers of energy services. Information programmes include publicity campaigns, energy audits, technical manuals and other forms of technical transfer systems, advisory services, training and education. In developing countries it is useful to note two types of information programmes which are gaining currency: energy audits in industry and advisory services (conservation centres).

Energy audits offer specific advice to consumers by initially doing some form of survey of current energy use to determine the potential for savings. A recent World Bank study states that "at the plant level, lack of information about the appropriate technical options and the absence of expertise in energy management often hamper conservation, as does the lack of energy-auditing capability . . ."** Canada was the pioneer of a

---

*   IEA, *op. cit.,* p. 120.
**   J. Gamba, D. Caplin and J. Mulck Luyse, *Industrial Energy Rationalization in Developing Countries,* World Bank, Baltimore, 1986, p. 8.

mobile, computerized energy audit service and recently the Soviet Union decided to purchase several "energy buses" from Canada. UNIDO also has a programme of mobile energy audits for a number of East European countries based on the Canadian and European Community experiences. Taiwan, South Korean and the Philippines also had comprehensive auditing services and Pakistan and India had more specific ones. In Africa, the Tanzanian Industrial Research and Development Organization has conducted a low-level audit programme for companies in Dar-Es-Salaam. In one series of industrial audits in Tanzania it was found that 8.6% of the fuel bill was made up of heat losses due to lack of insulation and that a further 6.2% of the fuel bill was energy losses due to maintenance problems.*

Several non-OECD countries have established energy conservation centres to provide a focal point for energy conservation activities, conduct publicity campaigns, provide technical information and co-ordinate training. Thailand has an Energy Conservation Centre within its National Energy Administration but is preparing a joint government-private sector energy management centre to be known as the Energy Conservation Center for Thailand (ECCT).**

Financial Incentives

In the Asian Development Bank study, 8 of the 9 countries provided incentives. Six offered loans; 4 offered tax concessions. Only Thailand had a negative incentive based on the size of automobile engines. In the ADB evaluation it was felt that many of the incentives were not "powerful enough to produce substantial results".***

For many countries, the option of providing incentives was not feasible because of lack of available government resources.

Regulations and Standards

As the IEA study concluded, "regulations and standards can be valuable to keep the long-term momentum and to reach special market segments (e.g. the residential sector which is the least price responsive end-use

---

*   UNEP, *Energy Conservation in Developing Countries,* Report of the Executive Director, Nairobi, March 1986, p.31.
**   UNDP/World Bank, *Thailand: Issues and Options in the Energy Sector,* Report No. 5793-TH, September 1985, p. 187.
***   ADB, *op.cit.,* p. 93.

sector, rented buildings and markets heavily influenced by style and advertising (e.g. automobiles). They ensure minimum levels of effort, are useful during periods of energy price fluctuations and should be reviewed periodically."*

The ADB study also showed that regulatory measures were also common amongst the countries studied. After the major oil price increases certain measures were put in place to curtail consumption. Those measures restricted business hours, external lights, air conditioning and so on. However, Korea, the Philippines and Taiwan also require labels on some appliances and mandatory efficiency standards for new buildings.** The Philippines have also required industrial plants to implement conservation programmes, mandatory energy data reporting and border regulation.***

Research, Development and Demonstration

The IEA study showed that the main contribution to promoting energy efficiency over the rest of the century is likely to be made by the commercialisation and diffusion of existing technologies rather than research and development into new technologies. This makes demonstration programmes particularly important. However, it also appreciates the need for further R&D to develop new technologies into the next century.

The demonstration of conservation technology as a means of technology transfer is of great importance. Much of the research and development needed is mainly for the application of existing technology. What is important is to find useful ways to transfer knowledge from the industrialized world to other countries.

Nevertheless, there is R&D undertaken in non-OECD countries. For example, in Brazil, the new energy conservation programme is

---

*   IEA, *op.cit.*, p. 153.
**   ADB, *op.cit.*, pp. 94-5.
***   UNEP, *op.cit. p.30.*

supporting the development of more efficient refrigerators, lighting products, heat pumps and control systems.*

Many countries need more energy-efficient wood stoves to reduce wood consumption and improve the environment. The World Bank has been active in developing new wood stoves ** as have been many individual countries. In some countries, such as India more efficient kerosene stoves have also been developed.

Exemplary Role of Governments

Governments use energy directly and face most of the same obstacles that confront other energy users. By trying to improve their use of energy, they not only manage their own resources well but set an example of good conservation practices.

Little information is available about governments implementing programmes to reduce their own energy use. The Cabinet in Thailand approved an action plan to improve energy efficiency in the government transport fleet but there has been "no serious attempt" as yet to implement it***.

The establishment of highly visible conservation centres do help but setting an example is also very important.

Effective involvement of various organisations which work on energy conservation

The IEA conservation study found that investment in energy conservation is enhanced by the involvement of many types of organisations:

---

*   H. Geller et al., *Electricity Conservation in Brazil: Potential and Progress,* May 1987. Paper presented at the NATO Advanced Study Institute on Demand-Side Management and Electricity End-Use Efficiency, Povoa do Varzim, Portugal, July 20-31, 1987, p.5.

**   For example see World Bank, *Test Results on Charcoal Stoves from Developing Countries,* December 1986.

***   UNDP/World Bank, Thailand, *op.cit.,* p. 194.

energy conservation service industries which provide equipment, services, and advice; supply industries such as utilities which provide incentives or advice; non-governmental service clubs or voluntary industry groups which provide services and share information; and governments at all levels encouraging conservation. What these organisations and companies provide is an infrastructure and the necessary techniques and technologies for consumers to take action. In some cases it is simply bringing together companies from a particular industrial sub-sector to discuss common problems related to improving energy efficiency. In other cases, it is involving utilities or other energy suppliers to encourage conservation, because these suppliers have the advantage of already dealing directly with consumers.

For example, in Brazil, the federal government established a national electricity conservation programme in 1985. The programme combines technology development and testing, market analysis, financial incentives, educational and training activities, regulatory measures and institutional development. The programme is based in Eletrobras, the electric utility, and also includes the participation of other utilities.

In many countries, international organisations — both governmental and non-governmental — are more active in encouraging energy conservation. These organisations include the development banks, both global and regional aid agencies, but also include specific organisations, such as the International Institute for Energy Conservation, based in the United States, which was formed in 1984 to "facilitate . . . the needed transfer of knowledge on energy efficiency from the industrialized world to the Third World."

Conclusions

Considerable energy conservation activity is being undertaken in many non-OECD countries. Understanding both how energy is used and how to imporve that use is very complex. Attention has to be given to what potential for efficiency improvements there is and how best to achieve all our part of that potential. Different policies and programmes can often produce similar results but at different costs and different unintended effects/benefits. It is difficult to generalize what is beneficial and what is not because conservation activities are so disaggregated.

IV. REGIONAL ENERGY CO-OPERATION IN LATIN AMERICA AND THE ASIA/PACIFIC REGION

A. Introduction

This chapter looks at two regions which have been very active in developing regional networks: Latin America and the Asia/Pacific. Both of these regions have had a variety of energy-related regional organisations, most of them, for more than a decade: some specifically for energy and some where energy is only one component; some that were policy oriented and some which were largely technical.

B. Latin America

Latin America generally refers to the countries of South America, Central America, Mexico and the Caribbean. The region has about 385 million people or about 8% of the world population and consumes about 5% of world primary energy. The largest country is Brazil with a population of over 136 million. Some of the small islands in the Caribbean have but a few thousand. The region includes 10 of the most indebted nations in the world. It also includes several major oil exporting countries.

Since the mid-1960s with the creation of the Mutual Assistance of the Latin American State Oil Enterprises (ARPEL), and the Commission for Regional Electrical Integration (CIER), Latin America has attempted to improve regional co-operation in the field of energy. After the first major oil price increase in the early 1970's, the Latin American Energy Organisation (OLADE) was formed.

Latin American Energy Organisation (OLADE)

OLADE was formed in 1973 and has its headquarters at Quito, Ecuador. Its member states — 26 in all — include energy producers and consumers, major net exporters and net importers. As it is a regional organisation, with all its members contiguous within the region, its ratifying document, the Lima Agreement, tailors its functions to the needs of a specific geographic area.

Broadly speaking, OLADE aims to serve as a focal point for the development of Latin American energy co-operation. Specifically, improved systems integration and increased regional energy trade are objectives to be pursued through promotion of policies supporting technical and developmental assistance, enhanced investment opportunities, and improved energy data exchanges to facilitate all of these goals. In addition to its continuing concern that Latin America develop an efficient and beneficial market for hydrocarbons, OLADE has concentrated upon, and succeeded in, promoting major international hydroelectric and geothermal projects. Considerable research work in the area of small new and renewable energy projects has also been carried out.

In November 1987, the XVII Meeting of Ministers gave strong support to the organisation and stressed the need for it to be "the foremost energy forum in Latin America and the Caribbean" and "to promote the active presence of OLADE at the international level". The Ministers stated that the priorities for 1987-1990 would be: energy planning, energy balances, rational use of energy, new and renewable sources of energy, technical-commercial exchange in the regional energy sector, and exchange of experiences in the area of energy financing.

Mutual Assistance of the Latin American State Oil Enterprises: (ARPEL)

ARPEL was formed in 1965 to allow state-owned oil companies to exchange information and technical assistance. Initially there were 8 state oil companies that were members. This has now grown to 16 including affiliates from 14 countries and three more are planning to join in the near future.* The organisation performs studies in such areas as expansion of the oil industry in Latin America and the expansion of oil industry service companies; the conservation of hydrocarbon resources; and the development, among its members, of commercial transactions. ARPEL regularly holds conferences, technical-scientific meetings and congresses. Only one state oil company per country is allowed to be a member, although others are allowed to be affiliates.

---

* Canada, through Petro-Canada International, is the only non-regional state oil company which is a member.

Commission for Regional Electrical Integration: (CIER)

The Commission was formed in 1964 on the initiative of the electrical authorities of Uruguay. The objective of the organisation is to promote and favour electrical integration in the region in all aspects such as: a) the best efficiency of organisations and enterprises from the member countries; b) exchange of information, experience and studies in the technical, economic, juridical and other fields, connected to the different activities of the organisations and enterprises; c) systematic exchange of personnel at all levels, among organisations and enterprises, and training of such personnel; d) technical assistance and co-operation among enterprises; e) projects with a regional concept, bearing in mind, especially, the possible establishment of international electrical interconnections; f) guidance and co-ordination of activities of common interest for the organisations and enterprises; g) co-ordinated purchase procedures and technical specifications; h) utilisation of technicians of the region for projects and studies, whenever possible.

CIER includes 10 South American countries: Argentina, Bolivia, Brazil, Colombia, Chile, Ecuador, Paraguay, Peru, Uruguay and Venezuela.

Economic Commission for Latin America (CEPAL)

Another regional organisation that has demonstrated an interest in energy is CEPAL, the Economic Commission for Latin America, which was created in 1948 by the United Nations to aid in the development of Latin America. CEPAL's basic mandate is to investigate and analyse the tendencies and problems of Latin American development, to advise the governments in the search for and application of solutions to these problems, and to train professionals in the socio-economic field.

During the mid- to late 1970s, CEPAL became interested in energy and began to participate in, as well as sponsor, activities that promoted regional energy co-operation in research, technical co-operation, training, and financing.

By 1983, CEPAL was very active in the energy area, having moved well beyond the research stage as an active promoter of co-operation, to some extent in co-operation with OLADE. CEPAL continues to monitor events in the region and conducts on-going research on energy-related issues.

Inter-American Development Bank (IDB)

The Inter-American Development Bank (IDB) was created in 1959 by 19 Latin American countries and the United States. It now has 44 members including 15 European countries (Austria, Belgium, Denmark, Finland, France, Germany, Italy, the Netherlands, Norway, Portugal, Spain, Sweden, Switzerland, the U.K. and Yugoslavia), Israel, and Japan. In 1985, the IDB completed twenty-five years of operation, during which it has channelled more than $28 billion in loans and technical co-operation to finance development projects in Latin America, representing a total investment of about $100 billion.

The non-regional members contribute both to the bank's capital, which is used to support loans on conventional terms, and to a "Fund for Special Operations", which provides concessional resources.

Of the Bank's loan portfolio, approximately 28% of loans has gone towards energy-related projects. Although the bulk of this has gone into projects for the expansion of electricity generating capacity, there also is money earmarked for less traditional projects to foster regional energy co-operation. Although the Bank's major thrust has been the conventional electricity sector, it has also been active in hydrocarbons, coal, and to a lesser extent in non-conventional energy projects in geothermal and small hydropower.

With respect to overall economic cooperation and development, the Bank has been one of the region's main promoters of integration — primarily through its loans and technical co-operation operations and through the Institute for Latin American Integration (INTAL). The Bank has promoted its interest in economic development through a number of seminars on economic challenges facing Latin America. Recent seminars have included the Central American Bank for Economic Integration (CABEI), and dealt with ways to overcome current economic problems and restore the pace of development. IDB undertook initiatives with the Latin American Free Trade Association (LAFTA) — now the Latin American Integration Assocation (LAIA) — and the Institute for Latin American Integration (INTAL) on ways of rejuvenating intra-Latin American trade. With the Board of the Cartagena Agreement's (JUNAC) participation, options and new approaches for regional development and integration were similarly examined.

By the end of 1986 the Bank had approved a total lending volume of $ 1,982 million for integration-related projects.

C. Asia/Pacific

The Asia/Pacific region has approximately 60% of the world's population. It comprises close to 40 countries from Japan on the east, to Pakistan in the west, Mongolia in the north and Australia and New Zealand in the south. There are four newly industrialising countries (Korea, Hong Kong, Singapore and Taiwan).

In terms of the distribution of energy resources, countries in the region have 3 types of situations:

— net energy exporting countries: e.g. Australia, Brunei, China, Indonesia and Malaysia;

— highly energy import dependent countries: e.g. Hong Kong, Japan, Korea, Singapore and Taiwan; and

— energy importing countries with significant domestic production: e.g. Bangladesh, India, New Zealand, Pakistan, the Philippines and Thailand.

The Economic and Social Commission for Asia and the Pacific (ESCAP)

The largest regional organisation in the Asia/Pacific is ESCAP, which is headquartered in Bangkok, Thailand. It has 31 regional member countries comprising 28 developing countries (including Iran), 3 developed countries in the region (Australia, Japan and New Zealand), and France, the Netherlands, the United Kingdom and the United States. ESCAP has been taking an active role in the region's energy development and co-operation since 1974, when a conference was held on the "Impact of the Oil Shock on ESCAP Member Countries" to seek a swift response to the first oil crisis of 1973.

The main features of ESCAP's approach to energy include a programme made up of 3 elements:

(i) Energy assessment and planning;

(ii) Accelerated development and use of new and renewable sources of energy (NRSE); and

(iii) Integrated investigation, development, conservation and efficient use of overall energy, with emphasis on conventional sources of energy.

Parts of this programme represent technical co-operation projects using extrabudgetary resources provided by the United Nations Development Programme (UNDP), the governments of Japan and Australia, and others.

Some of these technical co-operation projects such as The Regional Energy Development Programme (REDP), The Pacific Energy Development Programme (PEDP), and the Biomass/Wind-Energy Network are large, time-bound activities designed to achieve particular impacts within the framework of the overall programme.

The REDP is a programme for the developing countries in the region with the long-term objective to assist participating countries in the areas of planning and management of energy programmes, the efficient use of energy and the development of both conventional energy and new and renewable sources of energy. The programme was funded by the United Nations Development Programme (UNDP) for $3.5 million and implemented in collaboration with other United Nations organisations (such as UNIDO, FAO, UNESCO, ILO) and the Asian Development Bank. Currently, a second phase is being carried out, focussing on:

(a) energy assessment, planning and management, including energy pricing policy, energy conservation in small and medium scale industries, and energy planning and modelling;

(b) large scale systems such as natural gas, coal and trans-country power development; and

(c) rural energy systems and NRSE.

The PEDP aims at assisting Pacific island nations in securing energy and at facilitating a significant reduction in their reliance on imported petroleum products through improved capabilities in energy management and planning. The programme was also funded by UNDP.

The Regional Programme of Action on NRSE is a follow-up of the Nairobi Programme of Action on the development and utilization of NRSE. The Japanese government provided support for the establishment of a regional network on biomass, solar and wind energy (B-S-W Network). The Network undertakes such activies as the exchange of data on biomass, solar and wind sources, and dissemination of information on research, development and demonstration programmes.

Also under the ESCAP framework, energy issues have been dealt with in various contexts, such as the utilization of surplus agricultural residues as an energy source for productive activities in the agriculture programme, energy conservation in small and medium scale industries, and assistance in establishing a common framework for energy data collection and compilation.

The ASEAN Council for Petroleum (ASCOPE)

ASCOPE was established in 1975 with headquarters in Jakarta, Indonesia to foster co-operation among Association of South East Asian Nations (ASEAN) member countries in the following areas:

(i)   to promote active collaboration and mutual assistance in the development of the petroleum resources in the region through joint endeavours;

(ii) to collaborate in the efficient utilization of petroleum;

(iii) to provide assistance to each other in the form of training, and the use of research facilities and services in all phases of the petroleum industry;

(iv) to facilitate the exchange of information which will promote methodologies leading to successful achievements in the petroleum industry, and which may help in formulating policies within the industry;

(v)   to conduct petroleum conferences on a periodical basis;

(vi) to maintain close and beneficial co-operation with existing international and regional organisations with similar aims and purposes.

The 6 Member countries (Brunei, Indonesia, Malaysia, Philippines, Singapore and Thailand) are represented by the heads of state oil companies or national oil agencies.

Some of the highlights of the ASCOPE programmes are:

(i)   *Co-operation in Petroleum Exploration,* including geological information exchange and formulation of common regulations among the ASCOPE member countries in the field of petroleum industry operations (e.g. regulations regarding offshore exploration and production operations, environmental protection, pollution control, and marine transportation);

(ii) *Co-operation in Petroleum Production,* including information exchange regarding production techniques, equipment, exploration costs and exchange of information on the demand and supply of crude oil and petroleum products, conservation, and retail prices of products;

(iii) *Manpower Requirements and Training.* Some of the programmes in this area were organized in co-operation with industrialised countries such as the United States, Canada and Norway, and the UN's Committee for Co-ordination of Joint Prospecting in Asian Off-shore (CCOP);

(iv) *Sharing of Petroleum Products* in times of crisis or emergency. The ASCOPE member countries have adopted "The ASEAN Emergency Petroleum Sharing Scheme and the Supplementary Scheme" to protect member countries against critical shortages of crude oil and/or products for national and economic security; and

(v) *Cooperation in Systematizing the Storage, Collection, Dissemination, Transmission and Format of Petroleum Data.*

At the ASEAN Foreign Ministers Conference held on 24th June 1986 in Manila, Ministers signed two agreements in the area of energy: the ASEAN Petroleum Security Agreement and the Agreement on ASEAN Energy Co-operation. The first one formalised and expanded the already existing ASCOPE emergency oil and petroleum products sharing scheme into the ASEAN Emergency Petroleum Sharing Scheme so that it can be applied to both shortage and over-supply. The System works on the basis of the oil reporting system which gathers oil information once every three months. During the second oil crisis ASCOPE actually mobilised emergency sharing by allocating diesel oil from the Philippines to Thailand and crude oil from Malaysia and Indonesia to Thailand in the amounts of 5,000 b/d and 10,000 b/d respectively, in order to rectify shortfalls.

The second agreement also formalised a wide range of energy cooperation promoted by the ASCOPE in the past 10 years. Some of the main features of this agreement are training of engineers for exploration and refinery management, coordinated oil and gas exploration, development of alternative energy sources and promotion of energy conservation.

An ASEAN oil stock pile was proposed by Thailand in April 1985 at the ASEAN Energy Ministers Conference and is currently being studied by the member countries.

Asian Productivity Organisation (APO)

The APO was established in 1961 by several governments of the region to increase productivity and accelerate economic development by mutual cooperation. Today the APO has 16 member countries. The organisation is based in Tokyo.

The APO offers training projects in management and technology for development, research and surveys, consulting services, training manuals, audio-visual aids and various publications. The organisation has incorporated energy related projects into its training courses such as the Energy Management Course on Plant-Level Fuel Efficiency, Energy Policy for the Manufacturing Sector, Application of Alternative Energy Sources, Energy Conservation in the Paper and Pulp Industry, Energy Conservation in the Cement Industry, and Energy Management in Selected Asian Countries. The APO also undertakes energy-related studies, such as on energy utilization and energy conservation in the fertilizer industry.

Asia and Pacific Development Centre (APDC)

APDC is an intergovernmental organization which came into existence in 1982 under the aegis of ESCAP. Its headquarters are in Kuala Lumpur. It has 20 member countries in the region including 3 OECD countries: Australia, Japan and New Zealand. As the first phase of its activities, APDC selected four programme areas, one of which was energy planning and management.

The organisation concentrates on technology development and industrialisation after having completed the first round projects which incuded a project on producing a three volume textbook on integrated national energy planning. The Asian and Pacific Energy Planning Network or APENPLAN was created as a result of this project in March 1985 to help build or strengthen existing national capabilities in energy planning and management through the development of energy planning and management teams in countries of the region and also to establish close working associations with other international organisations with similar orientation. APDC has been organising a series of APENPLAN Workshops since September 1985 in various parts of the region to disseminate the techniques developed through the integrated national energy planning project.

The Forum on Minerals and Energy (MEF)

The Forum on Minerals and Energy (MEF) was established as part of the Pacific Economic Cooperation Conference (PECC) in April 1985. The PECC is a tripartite organistion of senior industry leaders, government officials and academic experts from the countries, regions and areas of the Western Pacific region and North America, which operates on an informal, non-exclusive and non-official basis:

— to examine key problems and issues influencing regional economic growth;

— to develop materials and findings helpful in enhancing Pacific economic cooperation and translating regional economic potential into performance;

— to stimulate efforts to find new mechanisms for solving common problems and reducing multilateral economic tensions among Pacific nations;

— to provide greater opportunity for regional interests to be pursued in other multilateral forums; and

— to promote public awareness and understanding of the increasing interdependence of the Pacific economies.

The general purpose of the MEF is to promote discussion and consultation among officials, industry leaders and independent researchers on minerals and energy issues of regional interest. The Forum seeks to identify potential sources of conflict or areas of co-operation, and will work to develop constructive suggestions on means by which governments and business organisations can facilitate the efficient development and trading of mineral and energy products within the region. Through the publication of its work and through its links with constituent members, the Forum promotes a more effective representation of Pacific interests in global forums concerned with minerals and energy issues.

The first official meeting of the MEF was held in Jakarta, Indonesia in July 1986. The detailed and broad ranging discussions highlighted the value of improvements in quality and exchange of information on prospective market developments and on changes in the policy environment. The second meeting of the MEF was held in Seoul, Korea, in October 1987.

Asian Development Bank(ADB)

The ADB established in 1966 is headquartered in Manila, Philippines and has 32 developing member countries and 14 industrialized member countries as its members.

The ADB attaches high priority to the effective development of energy facilities as well as to the planning and policy making for efficient management of the energy sector. The ADB's loan assistance for the energy sector projects adds up to about US$ 5.0 billion and accounts for about one quarter of its total assistance to its developing member countries.

Over the past few years, the ADB has been active in strengthening institutional planning capabilities in its developing member countries, including surveys such as the Asian Energy Survey and the Rural Electrification Survey, the compilation and publication of data such as the Asian Electric Power Utilities Data Book, Energy Indicators, and studies such as the Energy Policy Experiences of Developing Countries*.

In September 1987, the IEA and the ADB jointly organized an Energy Data Workshop in Tokyo for Asia-Pacific countries, the proceedings of which are to be published in mid-1988.

Other Organisations

The South Pacific Conference (SPC) supports small energy projects such as wind-powered pumps and solar lighting by way of its rural technology programme.

---

*   ADB, *Energy Policy Experiences of Developing Countries*, Manila, 1987.

The Interim Committee for Co-ordination of Investigations of the Lower Mekong Basin, Bangkok is engaged in a broad range of hydropower development investment studies and irrigation investigations.

V. THE FUTURE FOR NEW AND RENEWABLE SOURCES OF ENERGY

A. Introduction

After the Nairobi Conference on New and Renewable Sources of Energy (NRSE) in 1982, there was considerable enthusiasm in implementing the programme of action and undertaking directed and co-ordinated activities to encourage the development of new and renewable energy sources. With more than five years now past since the Conference it is useful to review the progress made and the status of the IEA's own activities in this field.

The interest in the use of renewable energy gained momentum after the first major oil price increase in the early 1970s. At that time, renewable energy technologies appeared attractive for industry because of the projected high cost of oil and because of their promise of modularity, rapid deployability and attendant financial risk reduction. They were also one of many options to improve energy security through less dependence on imported energy and had relatively benign environmental impacts. Thus, there was considerable RD&D activity and many programmes to develop and improve these technologies and make them commercially viable.

This momentum was still there when the Nairobi conference was held, but since then, the prices of conventional sources of energy have fallen and the attitude has changed. As stated in a recent study* by the IEA on this subject:

> "As a result of this experience, however, expectations concerning the pace of development and the contribution of renewable energies to energy supplies are now more realistic. This is partly because costs of conventional energies now seem unlikely to rise as far or as fast as was anticipated in the 1970s, and in fact, have recently fallen, affecting all alternatives. It is also because of a better assessment of the time needed for development and market penetration. These factors, combined with budgetary constraints, have weakened some governments' support and industry interest in developing alternatives to oil."

Nevertheless, in many developing countries the continuing and expanding use of renewable energy is very important. It is reflected in the energy balances where renewables contribute about 40% of energy needs in Latin America, 50% in the Asia-Pacific region, and 50% in Africa, mainly in the form of biomass (primarily fuelwood) which creates environmental problems. In some countries, renewables contribute more than 80% of total primary energy requirements. It has been estimated that over 2 billion people in developing countries "depend on fuelwood and charcoal for their household energy needs."**

B. Recent International Efforts to Encourage the Development of NRSE

Following the third session of the United Nations' Committee on the Development and Utilization of New and Renewable Sources of Energy which was held in June 1986, the UN General Assembly adopted a Resolution (A/RES/41/170) on 5th December 1986, which reaffirmed "the significance and importance of the Nairobi Programme of

---

*   IEA, *Renewable Sources of Energy*, OECD, Paris, 1987, p. 12.
** Paper presented to UN-ECE meeting by New and Renewable Sources of Energy
      Unit, Department of International Economic and Social Affairs

Action . . . as the basic framework for action by the international community and the United Nations system in that field," expressed concern about the slow rate of implementation of the programme of action and encouraged more intensive efforts.

Under the auspices of the regional UN organisations, there are Regional Consultative Meetings are held to bring together donors and recipients. For example, in November 1986, a regional meeting was held in Addis Ababa, Ethiopia. The meeting included 12 recipient countries, 14 donor countries and representatives of regional, subregional and international organisations. In that meeting the Deputy Executive Secretary of the UN-Economic Commission for Africa "emphasized the point that since the Nairobi Conference, NRSE have been advocated as a wide-spread solution to the energy problems of Africa. Unfortunately in some cases, the credibility of NRSE and renewable energy technologies has suffered from premature and excessive promotion of these technologies in Africa."*

Other bodies of the United Nations that are heavily involved in renewable energy include the Department of Technical Co-operation for Development which provides about $10 million annually for projects in the fields of energy planning, conservation, solar, wind, biomass geothermal energy and small hydropower. The United Nations Development Programme which provides funding for energy projects and, in conjunction with the World Bank, sponsors the Energy Sector Management Assessment Programme (ESMAP), which "offers policy makers a sharp focus on the key policy and investment decisions required for the sector, and a systematic ranking of priorities for action."** Increasingly their emphasis has been on rural issues and the development and utilization of new and renewable energy sources. Their country analyses have been valuable in assessing the renewables potential. The ESMAP programme has also prepared some cross-country analyses on specific energy issues including household energy use in Africa and technical studies such as one providing test results on charcoal stoves from developing countries.

In June-July 1987, the UN-Economic Commission for Europe sponsored a Symposium on the Status and Prospects of NRSE in the ECE

---

*   UN Economic Commission for Africa, ECA/NRD/RCM/6/86, p. 3.
**   UNDP/World Bank, *ESMAP*, April, 1987, p. 1.

Region. It confirmed the importance of renewable energy sources and acknowledged their potential medium- and long-term role. However, the share of NRSE in the energy balance in 2000 will increase only slightly and their are "still substantial gaps in resource assessments."[*] The recommendations are attached as an annex.

The Committee of Natural Resources of the Economic and Social Commission for Asia and the Pacific (ESCAP) has prepared an assessment of the contribution of NRSE.[**] While some technologies are very appropriate (such as small hydropower, wind energy conversion systems and solar photovoltaic technology (solar stoves) in remote areas), other technologies such as active solar heating and cooling of building, solar thermal electricity generation and the production of liquid and gaseous fuels from agricultural crops need "significant technological breakthroughs and/or very substantial cost rises in fossil fuels before they could be considered competitive."[***]

The International Labour Organisation (ILO) has given special attention to manpower and training assessments identified in the Nairobi Programme of Action. The ILO also has helped establish training facilities and its Turin Centre has developed training courses for renewable energy and has helped several countries establish activities related to the production of improved wood stoves. The ILO has also examined the social and economic impact of higher energy prices, new and renewable energy technologies, changing energy policies and the development of large-scale energy projects.[****]

The Food and Agriculture Organisation (FAO) has an active programme with four research networks and 14 working groups promoting the exchange of technical information and practical application of NRSE in rural areas of Europe and developing countries. This included workshops, reports, bulletins and state-of-the-art studies.

---

[*]   UN-ECE, *Report on Symposium...*, ENERGY/SEM.6/2, 4 August 1987.
[**]   ESCAP, *Assessment of the Contribution of New and Renewable Sources of Energy to Regional Energy Supply*, E/ESCP/NR.13/14, 27 August 1986.
[***]   *Ibid.*, p. i.
[****]   For a full review see, *ILO Activities in the Energy Sector with Special Reference to some Labour and Social Implications*, paper presented to UN-ECE Symposium, 29 June-3 July, 1987.

C. Related Activities of the International Energy Agency

After the Nairobi Conference the IEA prepared a document which outlined steps toward implementing the Nairobi programme of action.* That document outlined three of its activities "directed towards immediate facilitation of co-operative international efforts": energy assessment and planning; information flows, and education and training; and research, development and demonstration. The activities of the IEA since the Conference include:

(a) Energy assessment and planning

In 1984 the IEA published energy balances for 40 developing countries for the period 1971 to 1982. Currently it is updating the work and it is expected that energy balances to 1985 (and in some cases 1986) will be available in the latter half of 1988 for over 70 non-OECD countries.

(b) Information flows, and education and training

The IEA jointly sponsored a seminar on the rational use of energy in industry with the Latin American Energy Organisation, the Ministry of Energy and Mines of Peru and the Commission of the European Communities. While not the main topic, the field of alternative energy sources such as renewables was included in the seminar.

In September 1987 the IEA and the Asian Development Bank jointly sponsored a workshop in Tokyo on energy data. The purpose of the workshop was to report on the IEA's methodologies for data collection, evaluation and publication and to exchange views with developing countries on energy data procedures. Twelve developing countries from the Asia-Pacific region participated.

IEA has also made presentations at UN-Economic Commission for Europe seminars and has participated in seminars for the African Institute for Economic Development and Planning of the Economic Commission for Africa.

---

*   IEA, *New and Renewable Energy in the IEA: Steps Toward the Implementation of the Nairobi Programme of Action*, Paris, 1982.

(c) *Research, development and demonstration*

The main activity of the IEA in the area of renewables is in collaborative projects. A recent IEA publication, *A Ten-Year Review of Collaboration in Energy RD&D, 1976-1986,* details the projects that have been undertaken for geothermal, solar heating and cooling, small solar power systems, biomass conversion, ocean energy, wind energy and production of hydrogen from water. Some of the projects which have made good inroads include:

— innovative studies and analysis on meteorology contributing to the design of solar energy devices for space heating and cooling and the production of electric power;

— the world's first man-made geothermal energy source in New Mexico has yielded unique experience on drilling, fracturing and heat recovery techniques at great depths, and has revealed the feasibility of hot dry rock;

— the first major demonstration of wave power in the Sea of Japan and has led to a number of refinements to the original wave-breaking buoy;

— the multiple design solar power plant built in Almeria, Spain has revealed the limitations as well as the technical possibilities of solar electric power.

Altogether there have been 46 projects under 11 implementing agreements in renewable energies. Mexico has participated in one of the renewable projects (geothermal equipment testing).

(d) *The IEA study on Renewable Sources of Energy*

The other major work undertaken by the IEA has been the publication of a study on renewable energy sources which was completed in early-1987. It is useful to review the objectives and main findings of that report.

The study looks at 5 different renewable energy sources (solar, wind, biomass, geothermal and ocean energies) and makes an attempt to assess:

— the accomplishments and lessons of more than a decade of developments;

— the most promising areas of further RD&D and/or other government action to increase the development and utilisation of renewable energies; and

— the outlook for contributions to national energy supplies on an economic basis.

The combined technical and economic status of the technologies were classified into four stages:

— "Economic" technologies which are well developed and economically viable at least in some markets and locations; further market penetration will require technology refinements, mass production, and/or economies of scale;

— "Commercial-with-Incentives": technologies which are available in some markets but are competitive with the conventional technologies only with preferential treatments; these technologies still need further technology refinements, mass production and economies of scale;

— "Under-Development": technologies which need more R&D to improve efficiency, reliability or cost so as to become commercial;

— "Future-Technology": technologies which have not yet been technically proven, even though they are scientifically feasible.

Figure 5 also lists the various types of technologies under the four categories.

The study also points out that there are many biases which prevent these technologies from achieving significant market penetration in a timely fashion. These include the barriers common to any new technology which is being introduced into the market, such as a lack of standardization of systems; uniform approval practices; public understanding of the probable impacts or benefits; infrastructure and conventional financing. There are also other impediments: a tendency to require higher rates of return on investment in renewable than in conventional sources of energy; and utilities, which play a key role in the development of some renewable energy sources, sometimes have discouraged competition from renewable energy technologies.

The study concludes that remarkable progress has been made during the past twelve years to develop these new technologies. While they will

Figure 5 **Current Status of Renewable Energy Technologies**

Economic (in some locations)

Solar water heaters, replacing electricity, or with seasonal storage and for swimming pools
Solar industrial process heat with parabolic trough collectors or large flat-plate collectors
Residential passive solar heating designs and daylighting
Solar agricultural drying
Small remote photovoltaic systems
Small to medium wind systems
Direct biomass combustion
Anaerobic digestion (of some feedstocks)
Conventional geothermal technologies (dry and flashed steam power generation, high temperature hot water and low temperature heat)
Tidal systems *

Commercial-With-Incentives

Solar water and space heaters replacing natural gas or oil
Electricity generation with parabolic trough collectors
Non-residential passive solar heating and daylighting
Biomass liquid fuels (ethanol) from sugar and starch feedstocks
Binary cycle hydro-geothermal systems

Under Development

Solar space cooling (active and passive)
Solar thermal power systems (other than parabilic trough collectors)
Photovoltaic power systems
Large-sized wind systems
Biomass gasification
Hot dry rock geothermal
Geothermal total flow prime movers
Wave energy systems

Future Technologies

Photochemical and thermochemical conversion
Fast pyrolysis or direct liquefaction of biomass
Biochemical biomass conversion processes
Ocean thermal energy conversion systems
Geopressured geothermal
Geothermal magma

---

* Only one tidal system, which was built in 1968, is presently operating commercially (in France). All other tidal projects are either proposed or in a demonstration stage, and would be considered to be "Under Development" until built and proven "Economic" at a specific site under today's conditions.

Source: IEA, *Renewable Sources of Energy,* Paris, 1987

make an increasing contribution, even in a climate of substantial support for technology development, at least 30 years may be needed for them to achieve a significant market penetration.

As a follow-up to the study, the IEA is continuing to do more analysis on the economics of renewable energies.

D. Conclusions

As the activities in the international community have shown, there is still considerable attention being given to renewables. The resolution of the UN General Assembly and the work of the various UN bodies testify to the commitment to continue with the Nairobi Programme of Action.

The IEA continues to pursue its original objectives which include encouraging the development of alternative energy sources which can reduce the dependence on imported oil on an economic basis. For the IEA countries renewables will be playing an increasing, but still relatively small, role. It is especially important to those countries with remote, off-grid regions where conventional energy sources are very expensive.

The IEA's collaborative R&D projects have been particularly important and have made valuable contributions to the state-of-the-art.

Recommendations from UN-ECE Symposium on the Status and Prospects of NRSE in the ECE Region

29 June - 3 July, 1987

Recommendations

—   Development of a methodology for NRSE statistics.

—   Increased funding for national research centres, and grouping of research activities in each country within institutions dealing also with energy conservation.

—   Increased cooperation between research and industry. Participation by industry and states in finalizing research findings.

—   Incorporation of NRSE projects into global financing of the national sectoral programmes to which they will contribute.

—   Increasing emphasis to be placed on NRSE which have proved their technological maturity, economic viability and positive effect on the countries' economy.

—   Developing NRSE as a matter of necessity because of their:

    —   character as a national resource;

    —   contribution to industrial activity, and

    —   positive repercussions on advanced research capability.

— Removal of obstacles to the free penetration of NRSE on the international market.

— Continued development of local NRSE and relevant technology.

— Environmental implications, including pollution analysis (CO_2) and cost assessment, to be taken into account in energy policies.

— Order of priority for the development of NRSE to be established in each country.

— Strengthening of international collaboration on NRSE.

VI. ENERGY STATISTICS

The subject of Energy Statistics in non-OECD Countries is vast and complex but highly important to energy policy decisions of all countries, OECD or non-OECD. Since OECD energy statistics are used as a basis for comparison it is useful to consider the objectives underlying OECD statistics and their relevance for countries outside the OECD. Differences in objectives may often explain why non-OECD countries will not always need to compile statistics with the speed or in the detail found in OECD countries. On the other hand, a clearer definition of the basic objectives of energy statistics can often aid in strengthening statistical systems.

OECD countries require energy statistics systems to support activities which include the following:

— preparedness for oil emergencies;

— monitoring of energy markets;

— the promotion of energy efficiency;

— promotion of other energy sources as alternatives to oil;

— comprehensive review of national energy policies;

— study of the interaction of energy developments with overall economic developments;

— pursuit of energy research and development;

— understanding of the global context within which OECD energy country policies are developed;

— study of fiscal and financial aspects of energy developments;

— identification and resolution of issues related to energy and the environment;

— identification and removal of trade barriers between OECD countries; and

— provision of information to the energy and other industries of OECD countries.

While countries outside the OECD may share many of these objectives, they may not necessarily be concerned with all of them and may have concerns and objectives other than those of OECD countries. For example, information is required in the CPEs for planning purposes which is different in nature from information normally required in market economies. Again, many developing countries will be concerned about measuring the contribution of traditional fuels such as fuelwood and bagasse which are not widely used in most OECD countries.

It is outside the scope of the present report to examine the energy policy objectives of non-OECD countries in detail. However, it is clear that there are differences as well as similarities between the energy information requirements situation of different countries. Together the specific information needs of developing countries and CPE's, and certain similarities in the OECD countries information needs, point to ways in which energy statistical systems might be strengthened. The following sections look at the problem from the viewpoint of both economic structure and institutions together with their impact on the availability and management of energy statistics.

Organisation of Government and the Economy

A key question in the development of statistical systems is whether a country has a federal or centralised system of government. Federal countries are composed of states with various degrees of local autonomy.

This means that energy data systems of various kinds often exist at the state level rather than the national level. National data can sometimes be collated only from state data, which may vary in its specifications and degree of detail. Centralised countries will, on the other hand, produce most data at the national level. When regional data is produced in such countries it is likely to correspond to a national norm. Often, therefore, the degree of centralisation in a country may be more relevant than other characteristics such as degree of economic development, economic structure or extent of indigenous energy sources.

Another consideration is the extent of planning mechanisms. The existence of energy balances produced as part of a planning process does not guarantee the existence of strong energy data systems on a continuing basis. Often data will be assembled on an ad hoc basis for such purposes. Once plans have been formulated and agreed the energy data collection mechanism may not be used for several years until the plan is updated. There are, however, surprising contrasts among the CPE's. Energy information for Poland and Hungary compares well with that for many OECD countries. On the othe hand, the Soviet Union provides detailed and up-to-date data on energy supply but little on demand.

Many countries have legislation requiring public disclosure of various kinds of energy information, public or private sector enterprises or both. The legislation may apply to utilities which have to publish specified information on their activities as a condition of their operation, or less frequently, it may apply to industries which have to disclose their energy costs and consumption. In general, legislation of this kind is likely to be more common in OECD countries than in non-OECD countries.

Organisation of Energy Markets

There is considerable variety in the way in which energy markets are organised in different countries. In the case of oil, for example, there may be complete state monopoly at one or all phases from production to marketing, or it may be in the hands of the private sector. Oil balances are obtained in completely different ways in these two cases.

However, many countries have industry federations which have comprehensive statistical systems providing detailed coal, oil, gas or electricity balances. In such cases, the industry federation may be able to meet needs for national data efficiently, thus avoiding the creation of new data systems. Care must be taken to investigate that the federation provides 100% coverage of operations, e.g., an oil industry federation may exclude the petrochemical industry and/or oil traders, and/or regions. While such federations may play a role in some areas outside the OECD, notably in Latin America, there is a marked contrast between the active role they play in providing statistics in OECD countries and their minor role elsewhere.

Ministerial Responsibilities

When a single ministry is responsible for all energy sources (oil, coal, gas and nuclear) it is relatively easy to co-ordinate the data activities of the ministry and ensure coherence and completeness of the basic energy commodities, energy balances and overall balances. Dividing responsibility for energy among a number of ministries does complicate the task of energy statisticians. There are still many countries where, for example, oil and gas is the responsibility of one ministry, coal is the responsibility of another and the nuclear industry placed in yet another ministry, possibly that concerned with research and development. Many countries have resolved such problems successfully with coordinating groups.

Existing Data Systems

Energy data systems can produce energy statistics as a by-product or can be built to solely produce energy statistics. Sometimes the by-product type statistics can be of good quality. This applies, particularly, in the case of fiscal statistics concerning, for example, gasoline sales. On the other hand, surveys of industrial operations, for example, may produce better results when they are run by energy ministries rather than by central statistical offices.

Data Requirements of International Organisations and National Responses

Although not the main factor, requests from international bodies for information have been significant in promoting the efficiency and accuracy of national energy statistical systems. The effectiveness with which data needs of international organisations are met generally vary according to whether questionnaires are completed in ministries or by the energy industry (sometimes on behalf of ministries) or through national statistical offices. Some countries have in fact set up organisations for clearing and coordinating replies to international bodies. Once set up these coordinating organisations can be useful in meeting national needs for energy commodity balances and overall energy balances. Especially outside the OECD, moreover, non-Governmental bodies such as the World Energy Conference (WEC) and the International Union of Producers and Distributors of Electrical Energy (UNIPEDE) may be more successful in gathering information than inter-governmental statistical bodies. The IEA has had a strong influence on the collection and presentation of energy statistics outside the OECD by providing simple but thorough models for definitions, formats and conversion procedures.

Current IEA Work on Non-OECD Energy Statistics

Work is continuing in the Secretariat to provide a continuously updated data base of non-OECD energy statistics of sufficient detail and quality to support work on monitoring global energy developments, IEA forecasts of world energy supply and demand, and oil supply/demand developments.

The Secretariat supplements the UN 'World Energy Supplies' data base by material obtained directly from the countries concerned. The contacts made in the course of this work will often have a potential for information exchanges wider than those of energy supply/demand statistics alone.

The work on non-OECD countries is being pursued in close collaboration with other international organisations. A notable example of this was the joint workshop with the Asian Development Bank in Tokyo in September 1987. This resulted in a useful exchange of views and information with energy data experts for the information available on

South East Asian countries. Other organisations with which we are working closely on non-OECD countries include the ECE (Geneva), the Statistical Office of the United Nations, the World Energy Conference and OLADE. There have also been wide contacts with industry, especially with international oil companies.

The IEA non-OECD data base will eventually include about seventy countries in various degrees of detail. National data of heterogeneous nature is converted and/or adjusted to fit the IEA format of balances of individual energy sectors expressed in original units (The "Basic Energy Statistics") and then presented as overall energy balances. The data base will initially covers the period 1971 to 1986 where possible.

VII.   THE MIDDLE EAST SITUATION

Well over one-half of the planet's known, recoverable oil is found in the proven reserves* of the Middle East**. Some current estimates place this amount at over 400 billion barrels or 57% of the world total. With respect to actual supply, the Middle East's largest contribution to world oil requirements was 37% in 1974 and 1976. Middle East oil production was at its highest (around 22 mbd) over the period 1973-79 (see Table 31).

The producers of over 95% of Middle East oil are members of OPEC. Through the 1970s, these Middle East OPEC members also provided the majority share (about two-thirds) of aggregate OPEC oil production (see Table 32 and Figure 6), and they continue to produce the bulk of the Organisation's output.

---

*   Those quantities which geological and engineering information indicate with reasonable certainty can be recovered in the future from known reservoirs with existing technologies under current economic conditions.
**   The region bordered by Syria and Iraq in the north, Iran in the east, Oman and the Yemens on the south, and the Red Sea and Suez·Canal in the West.

Table 31 World Oil Production 1960-1986
(mbd)

	1960	1965	1970	1972	1973	1974
Middle East (mbd)	5.2	8.4	13.9	18.1	21.3	21.9
World inc. CPEs (mbd)	22.0	31.7	48.1	53.5	58.5	58.6
% Middle East	23.6	26.5	28.9	33.8	36.3	37.4

	1975	1976	1977	1978	1979	1980
Middle East (mbd)	19.7	22.4	22.5	21.4	21.9	18.8
World inc. CPEs (mbd)	55.7	60.1	62.6	63.1	65.8	62.7
% Middle East	35.4	37.2	36.0	34.0	33.3	29.9

	1981	1982	1983	1984	1985	1986
Middle East (mbd)	16.0	13.3	12.1	11.9	10.9	13.0
World inc. CPEs (mbd)	59.4	57.1	56.7	58.1	57.6	60.2
% Middle East	27.0	23.3	21.4	20.4	18.9	21.7

Source: BP Statistical Review of World Energy.

Figure 6 OPEC Oil Production

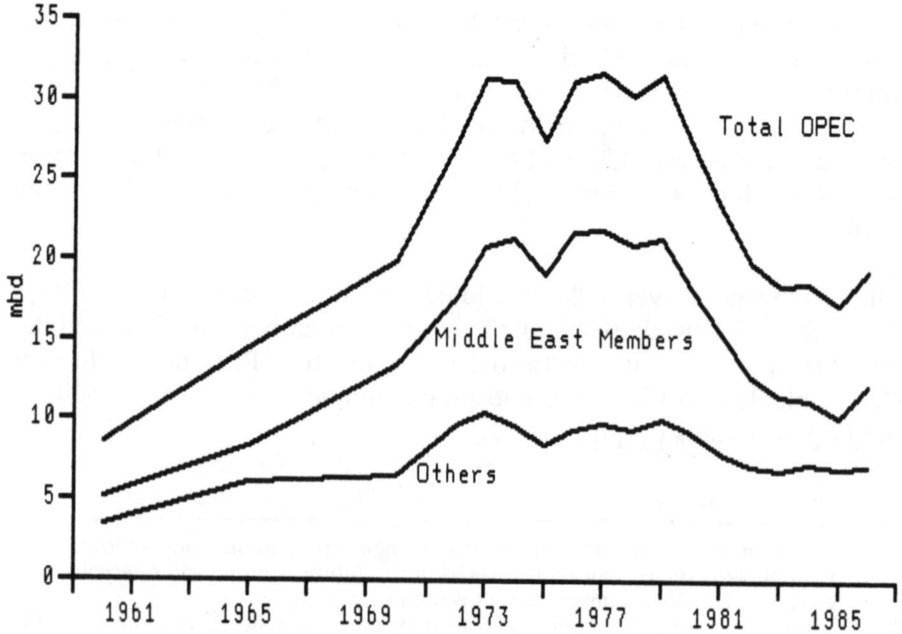

Sources: IEA database, BP Statistical Review of World Energy.

Table 32 **OPEC Oil Production 1960-1986**
(mbd)

	1960	1965	1970	1972	1973	1974
Middle East members(mbd)	5.2	8.3	13.4	17.6	20.8	21.4
Others	3.4	6.1	6.5	9.7	10.5	9.7
Total OPEC (mbd)	8.6	14.4	19.9	27.3	31.3	31.1
% Middle East	60	58	67	64	67	70

	1975	1976	1977	1978	1979	1980
Middle East members(mbd)	19.1	21.7	21.9	20.9	21.4	18.3
Others	8.4	9.4	9.8	9.4	10.1	9.2
Total OPEC (mbd)	27.5	31.1	31.7	30.3	31.5	27.5
% Middle East	69	70	69	69	68	67

	1981	1982	1983	1984	1985	1986
Middle East members (mbd)	15.5	12.7	11.5	11.2	10.2	12.2
Others	7.9	7.2	6.9	7.3	7.0	7.2
Total OPEC	23.4	19.9	18.4	18.5	17.2	19.4
% Middle East	66	64	63	61	59	63

Source: BP Statistical Review of World Energy.

The role of Middle East oil in meeting world needs decreased considerably during the 1980s, almost halving by 1985 as the demand for oil from OPEC countries diminished with weaker demand from OECD countries, and rising supplies from non-OPEC developing countries, and the North Sea (see Figure 7).

R/P ratio for Middle East oil (85 years) is much greater than the average (17.5 years) for other regions, and it is likely that a large proportion of future world needs will still come from there, as it has at times in the past.

Events in the Middle East have often affected oil availability:

1951-1953   cessation of oil company liftings from Iran during the Mossadegh era

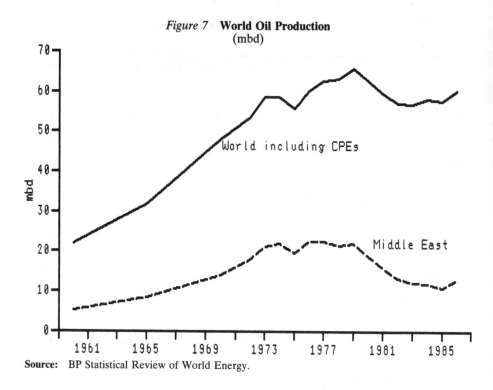

Figure 7 **World Oil Production**
(mbd)

Source: BP Statistical Review of World Energy.

1956	Suez crisis and closure of canal
1958	Iraqi revolution
1967	Arab-Israel war
1973	Arab-Israeli War; OAPEC oil embargo
1978-1979	Iranian revolution
1980	Beginning of Iraq-Iran War.

In addition to the specific events noted above, other long-running problems also exist with respect to Israel, Lebanon, North and South Yemen, and just outside the Middle East, in Afghanistan and Ethiopia. These problems, too, can affect the Middle East's petroleum industry.

At various times in the past, Middle East countries have set production ceilings for resource conservation reasons (e.g. Saudi Arabia 8.5 mbd and Kuwait 1.5 mbd in the mid-1970's, and Iran 3 mbd in 1979/80 after the revolution). Table 34 incorporates average production levels in 1987, maximum historical production levels, and current estimated sustainable production capacity.

Table 33 **Reserves/Production (R/P) Ratios**

	Proven Recoverable Reserves at end 1986 (bn bbls)	1986 Production (bn bbls)	R/P Ratio (years)
Middle East	402.0	4.76	85.5
Developing Countries			
Asia/Pacific	17.1	1.02	16.7
Latin America	88.9	2.43	37.7
Africa	55.2	1.91	29.3
Sub-total non-Middle East Developing Countries	161.2	5.36	30.1
Centrally Planned Economies (CPEs)			
China	18.4	0.96	18.5
Soviet Union	59.0	4.57	13.1
Others	1.9	0.17	11.0
Sub-total CPEs	79.3	5.70	14.0
OECD			
US	32.5	3.74	8.5
Canada	7.9	0.66	12.3
Norway	10.5	0.33	31.2
UK	5.3	0.97	5.5
Australia	1.7	0.27	6.8
Others	2.7	0.19	14.0
Sub-total: OECD	60.6	6.16	9.85
Sub-Total: World excluding Middle East	301.1	17.21	17.5
World	703.1	21.97	32.5

Note: Reserves of tar sands and shale are excluded.

Source: BP Statistical Review of World Energy.

For the OECD countries in the Asia-Pacific region, the proportion of their oil imports emanating from the Middle East countries still exceeds 50%. The trends for OECD regions are in Table 35:

A similar situation applies in the non-OECD regions, which generally rely on Middle East oil for less than 50% of their needs. The exception is the Asia-Pacific region. In 1985, non-OECD South East Asia took 61% of its oil imports from the Middle East, while South Asia took 78%.

Table 34 Crude/NGL Production Levels
(mbd)

	Average 1987 Production Levels	Maximum Historical Production Since End 1978*	Current Estimated Sustainable Production Capacity
Bahrain	0.04	0.05	0.05
Iran	2.26	4.1	3.0
Iraq	2.09	3.7	3.0
Kuwait	1.10	2.6	2.0
Neutral Zone	0.37	0.6	0.6
Oman	0.57	0.3	0.6
Qatar	0.28	0.5	0.6
Saudi Arabia	4.04	10.5	8.0
Syria	0.22	0.23	0.25
United Arab Emirates	1.49	1.9	2.2
Total Crude	12.46	24.48	20.3
Total NGL	0.80		
NGL and Crude	13.26		

* Average daily output for one month period.

Note: Maximum sustainable capacity does not reflect limitations imposed by hostilities, production ceilings, or pipelines.

Sources: IEA Secretariat, International Energy Statistical Review.

Table 35 Share of Middle East Oil in Total OECD Imports
(%)

	1978	1979	1980	1981	1982	1983	1984	1985	1986	1987
North America	31	28	26	24	15	11	10	7	15	19
OECD Pacific	74	72	70	66	66	64	64	63	62	61
OECD Europe	48	52	48	44	34	26	24	21	26	26

Source: Secretariat Estimates.

Oil Supply by Sea from the Gulf*

Table 36 shows that as a consequence of both OPEC's lower share of the world oil market, and the increasing amount of Saudi and Iraqi oil being piped either to the Mediterranean Sea (via Lebanon, Syria, or Turkey),

* The Gulf countries are defined as those Middle East countries with a coast-line in the Persian/Arabian Gulf.

or to the Red Sea (Yanbu), the proportion of internationally traded oil* passing out of the Gulf by sea declined substantially up to 1985, but it reversed in 1986.

Table 36 **International Oil Market Shares**
(%)

	1973	1978	1979	1980	1981	1982	1983	1984	1985	1986
OPEC share of world crude/NGL market outside CPEs	64	59	60	56	49	43	41	40	38	41
% of Internationally Traded Oil exported from Gulf by sea	57	60	56	53	48	38	37	34	29	32
% of Internationally Traded Oil being exported from Gulf countries by pipeline or truck	7	1	. 4	5	7	18	13	16	23	21

Source: Secretariat Estimates.

The share of Gulf exports by sea coming to OECD countries declined slightly from 1978 up to 1985, and then rose in 1986 as prices fell:

	1978	1979	1980	1981	1982	1983	1984	1985	1986
Share of Total Gulf Exports by Sea to OECD countries (%)	74	73	73	73	72	68	68	65	70

Up until 1986, the share of North America and OECD Europe in OECD country receipts of Gulf exports by sea was progressively reduced as the result of rising North Sea production, and the increased availability of Middle East pipeline oil arriving at the Mediterranean Sea. A rising share was thus directed towards OECD Pacific countries.

---

* The definition of "internationally traded oil" includes CPE gross exports.

Table 37   **Gulf Oil Exports by Sea to OECD Countries**
(%)

	1978	1979	1980	1981	1982	1983	1984	1985	1986	1987
North America	19	17	16	15	10	9	9	7	16	21
OECD Pacific	26	27	28	29	36	42	44	46	38	38
OECD Europe	55	56	56	57	54	49	47	47	46	41
	100	100	100	100	100	100	100	100	100	100

Source: Secretariat Estimates.

Recent expansions to export pipeline capacities were as follows:

— the looping of the Saudi Petroline to Yanbu to take capacity to 3.2 mbd from 1.85 mbd (completed in early 1987);

— the looping of the Iraqi line to Turkey to take capacity to 1.55 mbd from 1.05 mbd (completed in late 1987).

However, 3 major lines are closed at present:

— the Saudi Tapline to Lebanon − 0.5 mbd capacity, closed since February 1975;

— the Iraqi line to Lebanon − 0.5 mbd capacity, closed since April 1976;

— · the Iraqi line to Syria − 0.7 mbd capacity, closed since April 1982.

The extent to which they may be used again - for technical, practical, or economic reasons - is questionable.

Further pipeline capacity may be installed by:

— Iraq (Phase II of the Iraqi line via Saudi Arabia with a planned capacity of 1.6 mbd to a port south of Yanbu);

— Iran (to its south-eastern coast-line, and/or to the Black Sea in conjunction with the Soviet Union, and/or through to the coastline of Turkey);

— possible further expansion of the Iraqi line to Turkey;

— Kuwait (through to a Red Sea terminal).

Under circumstances of gradually rising oil prices, the proportion of internationally traded oil leaving the Gulf by sea could well remain at less than 25%, with about another 25% being exported by pipeline. Total dependence on Gulf sources in internationally traded oil could return to over 50% by the early 1990s, i.e. to levels which pertained over the period 1960-81. It may be that level itself, rather than a differentiation between ships and pipelines, is in the end the most important. The attacks on tankers since 1983 have so far caused little interruption to oil flows. On the other hand, there is always the possibility that oil exports by pipeline could be interrupted by serious damage to one of the key pumping stations.

Supply/demand logistics mean that the Asia-Pacific region is likely to remain the dominant recipient of oil leaving the Gulf by sea, whereas European imports of Middle East and Gulf oil (particularly those destined for Mediterranean countries) are more and more likely to be those transitting pipelines to the Red Sea or Mediterranean ports. Demand from North America for Gulf oil has traditionally tended to be met through liftings directly from Gulf ports.

Conclusions

With the construction of several new export pipelines in recent years, together with those projected, the shipment of oil from the Gulf by sea would not be as important a factor to oil security as it was up until the early 1980s. Dependence on Gulf oil in internationally traded oil by the early 1990's could rise to over 50%, as it was over the period 1960-81. This factor, rather than trade through the Gulf itself, is more likely to become a dominant economic factor.

Nevertheless, political events could still produce situations impeding shipment in the Gulf for periods of time, and affecting quantities of oil, which would threaten the energy security of oil importing countries, require then existing supply patterns to be adjusted appropriately to minimise the impacts.

GLOSSARY OF TERMS

bd	Barrels per day
bcm	Billion cubic metres
CPEs	Centrally Planned Economies: Albania, Bulgaria, China, Cuba, Czechoslovakia, German Democratic Republic, Hungary, Kampuchea, Laos, Mongolia, North Korea, Poland, Romania, USSR, Vietnam, Yugoslavia
GDP	Gross Domestic Product
kcal	kilocalories (10^3 calories)
kWh	kilowatt-hours
mbd	million barrels per day
Mtoe	million tonnes of oil equivalent
mt	metric ton = tonne
Mmt	million (10^6) metric tons
MW/MWe	Megawatt (10^6 watt) of electricity
NGL	Natural gas liquids: including ethane, propane, butane, pentane and the heavier hydrocarbons
OIDCs	oil importing developing countries
OPEC	Organisation of Petroleum Exporting Countries: Algeria, Ecuador, Gabon, Indonesia, Iraq, Iran, Kuwait, Libya, Nigeria, Qatar, Saudi Arabia, the United Arab Emirates and Venezuela
OXDCs	oil exporting developing countries
R/P	reserve/production ratio
TPER	Total primary energy requirements

WHERE TO OBTAIN OECD PUBLICATIONS
OÙ OBTENIR LES PUBLICATIONS DE L'OCDE

ARGENTINA - ARGENTINE
Carlos Hirsch S.R.L.,
Florida 165, 4º Piso,
(Galeria Guemes) 1333 Buenos Aires
Tel. 33.1787.2391 y 30.7122

AUSTRALIA - AUSTRALIE
D.A. Book (Aust.) Pty. Ltd.
11-13 Station Street (P.O. Box 163)
Mitcham, Vic. 3132   Tel. (03) 873 4411

AUSTRIA - AUTRICHE
OECD Publications and Information Centre,
4 Simrockstrasse,
5300 Bonn (Germany)   Tel. (0228) 21.60.45
Gerold & Co., Graben 31, Wien 1 Tel. 52.22.35

BELGIUM - BELGIQUE
Jean de Lannoy,
Avenue du Roi 202
B-1060 Bruxelles   Tel. (02) 538.51.69

CANADA
Renouf Publishing Company Ltd/
Éditions Renouf Ltée,
1294 Algoma Road, Ottawa, Ont. K1B 3W8
Tel: (613) 741-4333
Toll Free/Sans Frais:
Ontario, Quebec, Maritimes:
1-800-267-1805
Western Canada, Newfoundland:
1-800-267-1826
Stores/Magasins:
61 rue Sparks St., Ottawa, Ont. K1P 5A6
Tel: (613) 238-8985
211 rue Yonge St., Toronto, Ont. M5B 1M4
Tel: (416) 363-3171
Federal Publications Inc.,
301-303 King St. W.,
Toronto, Ont. M5V 1J5
Tel. (416)581-1552
Les Éditions la Liberté inc.,
3020 Chemin Sainte-Foy,
Sainte-Foy, P.Q. GIX 3V6,
Tel. (418)658-3763

DENMARK - DANEMARK
Munksgaard Export and Subscription Service
35, Nørre Søgade, DK-1370 København K
Tel. +45.1.12.85.70

FINLAND - FINLANDE
Akateeminen Kirjakauppa,
Keskuskatu 1, 00100 Helsinki 10   Tel. 0.12141

FRANCE
OCDE/OECD
Mail Orders/Commandes par correspondance :
2, rue André-Pascal,
75775 Paris Cedex 16
Tel. (1) 45.24.82.00
Bookshop/Librairie : 33, rue Octave-Feuillet
75016 Paris
Tel. (1) 45.24.81.67 or/ou (1) 45.24.81.81
Librairie de l'Université,
12a, rue Nazareth,
13602 Aix-en-Provence   Tel. 42.26.18.08

GERMANY - ALLEMAGNE
OECD Publications and Information Centre,
4 Simrockstrasse,
5300 Bonn   Tel. (0228) 21.60.45

GREECE - GRÈCE
Librairie Kauffmann,
28, rue du Stade, 105 64 Athens Tel. 322.21.60

HONG KONG
Government Information Services,
Publications (Sales) Office,
Information Services Department
No. 1, Battery Path, Central

ICELAND - ISLANDE
Snæbjörn Jónsson & Co., h.f.,
Hafnarstræti 4 & 9,
P.O.B. 1131 - Reykjavik
Tel. 13133/14281/11936

INDIA - INDE
Oxford Book and Stationery Co.,
Scindia House, New Delhi 110001
Tel. 331.5896/5308
17 Park St., Calcutta 700016   Tel. 240832

INDONESIA - INDONÉSIE
Pdii-Lipi, P.O. Box 3065/JKT.Jakarta
Tel. 583467

IRELAND - IRLANDE
TDC Publishers - Library Suppliers,
12 North Frederick Street, Dublin 1
Tel. 744835-749677

ITALY - ITALIE
Libreria Commissionaria Sansoni,
Via Lamarmora 45, 50121 Firenze
Tel. 579751/584468
Via Bartolini 29, 20155 Milano   Tel. 365083
La diffusione delle pubblicazioni OCSE viene
assicurata dalle principali librerie ed anche da :
Editrice e Libreria Herder,
Piazza Montecitorio 120, 00186 Roma
Tel. 6794628
Libreria Hœpli,
Via Hœpli 5, 20121 Milano   Tel. 865446
Libreria Scientifica
Dott. Lucio de Biasio "Aeiou"
Via Meravigli 16, 20123 Milano   Tel. 807679

JAPAN - JAPON
OECD Publications and Information Centre,
Landic Akasaka Bldg., 2-3-4 Akasaka,
Minato-ku, Tokyo 107   Tel. 586.2016

KOREA - CORÉE
Kyobo Book Centre Co. Ltd.
P.O.Box: Kwang Hwa Moon 1658,
Seoul   Tel. (REP) 730.78.91

LEBANON - LIBAN
Documenta Scientifica/Redico,
Edison Building, Bliss St.,
P.O.B. 5641, Beirut   Tel. 354429-344425

MALAYSIA/SINGAPORE -
MALAISIE/SINGAPOUR
University of Malaya Co-operative Bookshop
Ltd.,
7 Lrg 51A/227A, Petaling Jaya
Malaysia   Tel. 7565000/7565425
Information Publications Pte Ltd
Pei-Fu Industrial Building,
24 New Industrial Road No. 02-06
Singapore 1953   Tel. 2831786, 2831798

NETHERLANDS - PAYS-BAS
SDU Uitgeverij
Christoffel Plantijnstraat 2
Postbus 20014
2500 EA's-Gravenhage   Tel. 070-789911
Voor bestellingen: Tel. 070-789880

NEW ZEALAND - NOUVELLE-ZÉLANDE
Government Printing Office Bookshops:
Auckland: Retail Bookshop, 25 Rutland Stseet,
Mail Orders, 85 Beach Road
Private Bag C.P.O.
Hamilton: Retail: Ward Street,
Mail Orders, P.O. Box 857
Wellington: Retail, Mulgrave Street, (Head
Office)
Cubacade World Trade Centre,
Mail Orders, Private Bag
Christchurch: Retail, 159 Hereford Street,
Mail Orders, Private Bag
Dunedin: Retail, Princes Street,
Mail Orders, P.O. Box 1104

NORWAY - NORVÈGE
Tanum-Karl Johan
Karl Johans gate 43, Oslo 1
PB 1177 Sentrum, 0107 Oslo 1Tel. (02) 42.93.10

PAKISTAN
Mirza Book Agency
65 Shahrah Quaid-E-Azam, Lahore 3 Tel. 66839

PHILIPPINES
I.J. Sagun Enterprises, Inc.
P.O. Box 4322 CPO Manila
Tel. 695-1946, 922-9495

PORTUGAL
Livraria Portugal,
Rua do Carmo 70-74,
1117 Lisboa Codex   Tel. 360582/3

SINGAPORE/MALAYSIA -
SINGAPOUR/MALAISIE
See "Malaysia/Singapor". Voir
« Malaisie/Singapour»

SPAIN - ESPAGNE
Mundi-Prensa Libros, S.A.,
Castelló 37, Apartado 1223, Madrid-28001
Tel. 431.33.99
Libreria Bosch, Ronda Universidad 11,
Barcelona 7   Tel. 317.53.08/317.53.58

SWEDEN - SUÈDE
AB CE Fritzes Kungl. Hovbokhandel,
Box 16356, S 103 27 STH,
Regeringsgatan 12,
DS Stockholm   , Tel. (08) 23.89.00
Subscription Agency/Abonnements:
Wennergren-Williams AB,
Box 30004, S104 25 Stockholm Tel. (08)54.12.00

SWITZERLAND - SUISSE
OECD Publications and Information Centre,
4 Simrockstrasse,
5300 Bonn (Germany)   Tel. (0228) 21.60.45
Librairie Payot,
6 rue Grenus, 1211 Genève 11
Tel. (022) 31.89.50
United Nations Bookshop/Librairie des Nations-
Unies
Palais des Nations,
1211 - Geneva 10
Tel. 022-34-60-11 (ext. 48 72)

TAIWAN - FORMOSE
Good Faith Worldwide Int'l Co., Ltd.
9th floor, No. 118, Sec.2
Chung Hsiao E. Road
Taipei   Tel. 391.7396/391.7397

THAILAND - THAILANDE
Suksit Siam Co., Ltd., 1715 Rama IV Rd.,
Samyam Bangkok 5   Tel. 2511630
INDEX Book Promotion & Service Ltd.
59/6 Soi Lang Suan, Ploenchit Road
Patjumamwan, Bangkok 10500
Tel. 250-1919, 252-1066

TURKEY - TURQUIE
Kültur Yayinlari Is-Türk Ltd. Sti.
Atatürk Bulvari No: 191/Kat. 21
Kavaklidere/Ankara   Tel. 25.07.60
Dolmabahce Cad. No: 29
Besiktas/Istanbul   Tel. 160.71.88

UNITED KINGDOM - ROYAUME-UNI
H.M. Stationery Office,
Postal orders only: (01)211-5656
P.O.B. 276, London SW8 5DT
Telephone orders: (01) 622.3316, or
Personal callers:
49 High Holborn, London WC1V 6HB
Branches at: Belfast, Birmingham,
Bristol, Edinburgh, Manchester

UNITED STATES - ÉTATS-UNIS
OECD Publications and Information Centre,
2001 L Street, N.W., Suite 700,
Washington, D.C. 20036 - 4095
Tel. (202) 785.6323

VENEZUELA
Libreria del Este,
Avda F. Miranda 52, Aptdo. 60337,
Edificio Galipan, Caracas 106
Tel. 951.17.05/951.23.07/951.12.97

YUGOSLAVIA - YOUGOSLAVIE
Jugoslovenska Knjiga, Knez Mihajlova 2,
P.O.B. 36, Beograd   Tel. 621.992

Orders and inquiries from countries where
Distributors have not yet been appointed should be
sent to:
OECD, Publications Service, 2, rue André-Pascal,
75775 PARIS CEDEX 16.

Les commandes provenant de pays où l'OCDE n'a
pas encore désigné de distributeur doivent être
adressées à :
OCDE, Service des Publications. 2, rue André-
Pascal, 75775 PARIS CEDEX 16.

71784-05-1988

OECD PUBLICATIONS, 2, rue André-Pascal, 75775 PARIS CEDEX 16 - No. 44443 1988
PRINTED IN FRANCE
(61 88 08 1) ISBN 92-64-13115-9